Technology of Knit Spacer Fabrics

Technology of Knit Spacer Fabrics

Dr. N. Gokarneshan
Dr. K. M. Pachaiyappan
Dr. B. Senthil Kumar
V. Mahalakshmi
S. Suganthi
P. Gracy

WOODHEAD PUBLISHING INDIA PVT LTD

New Delhi, India

Published by Woodhead Publishing India Pvt. Ltd.
Woodhead Publishing India Pvt. Ltd.,
303, Vardaan House, 7/28, Ansari Road,
Daryaganj, New Delhi - 110002, India
www.woodheadpublishingindia.com

First published 2020, Woodhead Publishing India Pvt. Ltd.
© Woodhead Publishing India Pvt. Ltd., 2020

Woodhead Publishing India Pvt. Ltd. ISBN: 978-93-88320-33-7
Woodhead Publishing India Pvt. Ltd. e-ISBN: 978-93-88320-34-4

Typeset by T-soft Solutions

Printed and bound in India by Replika Press Pvt. Ltd.

Contents

Preface

Knit spacer fabrics fall under the category of technical textiles. They find a number of functional applications, most of which is in automotive and some in medical textile applications. Knit spacer fabrics are of three types, namely, warp knit, weft knit and flat knit. Of these, the warp knit are most commonly used in technical applications. The book deals with all the three types of fabrics, their design aspects and their areas of applications. A good deal of research has been done during the past decade relating to spacer fabrics. Though woven and nonwoven spacer fabrics are used in technical applications, knit spacer fabrics are most commonly used. The first chapter deals with the fundamental aspects of knit spacer fabrics. Second chapter deals with comparison of knit spacer fabrics. The third chapter deals with stab resistance of warp knit spacer fabrics suitable for body armor and compared with existing types. The fourth chapter highlights the development of warp knit spacer fabrics for cushioning applications and compared with PU foam. Fifth chapter deals with warp knit spacer fabrics produced for protection of human body against impact, in hemispherical form. The sixth chapter is concerned with the study of spherical compression behavior of warp knit spacer fabrics. The seventh chapter describes the analysis of the compression behavior of warp knit spacer fabrics for evaluation of suitability in cushioning applications. The eighth chapter deals with the assessment of various types of warp knit structures relating to impact behavior and damage characteristic. The ninth chapter analyzes the impact compressive behavior of warp knit spacer fabrics for evaluation of suitability in protective applications. The tenth chapter deals with the effect of fabric parameters on porosity and capillarity of weft knit spacer fabrics. The eleventh chapter is concerned with the sound absorption behavior of warp and weft knit spacer fabrics considering the influences of various fabric layers and arrangement sequences on the noise absorption coefficient. Twelfth chapter deals with quantitative study on warp and weft knit spacer fabrics with regard to different fabric characteristics and enable to select spacer fabric based on end use application. The thirteenth chapter concerns with use of innovative flat knitting technique for production of newer type of 3D spacer fabric in light weight composite applications. The fourteenth chapter deals with warp knit spacer fabrics intended for shoe insole applications, and wherein permeability and conductivity properties have been studied. The fifteenth chapter is concerned with development of weft knit spacer fabrics intended for pressure ulcer prevention. The sixteenth chapter deals with development of an artificial neural network algorithm to predict the heat and moisture transfer properties of warp knit spacer fabrics. The seventeenth chapter deals with influences of test boundary conditions and sample size on the compression stress strain characteristic of a typical spacer fabric intended as a cushioning material for human body protection. The eighteenth and last chapter is devoted to assessing suitability of warp knit spacer fabrics for automotive applications, particularly car seats. I wish to express my heartfelt gratitude to our beloved Chairman, Sri S. V. Balasubramaniam, Trustee, Sri M. P. Vijayakumar and Principal, Dr. C. Palanisamy for their moral support.

I do hope that the readers would find the book useful for their research work. It is mainly intended to research scholars and research workers in the field to stimulate future research in knit spacer fabrics. Suggestions for improvement of the quality of the book is most welcome.

Dr. N. Gokarneshan

Dean in charge,
Dept. of Costume Design and
Fashion,
Dr. SNS Rajalakshmi College
of Arts and Science,
Coimbatore, Tamil Nadu, India

Chapter 1

Introduction

Spacer fabric is a 3D knitted fabric consisting of two separate knitted substrates which are joined together or kept apart by spacer yarns (Figure 1) [1].

First layer – Hydrophillic nature

Second layer – hydroscopic nature

Spacer layer – mono or multi filament

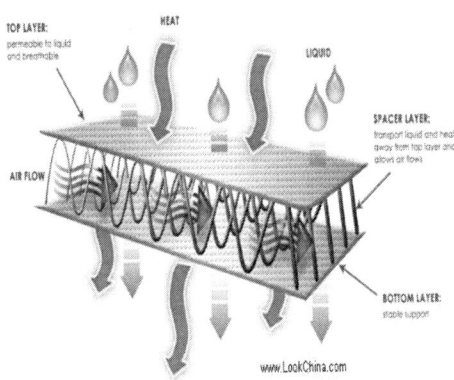

Figure 1 – Three layers of spacer fabric [1]

This 3D fabric is comprised of an initial layer for moisture release, an interior layer for air flow, and a third outer layer for heat dissipation.

According to the end uses the spacer ends of mono filaments may be polyester, polyamide or polypropylene. These fabrics are designed for air-flow and cushioning.

The middle to create two separate fabrics(spacer fabrics) are essentially pile fabrics that have not been cut consisting of two layers of fabric separated by yarns at a 90° angle.

1.1   Construction

Spacer fabrics (3D fabrics) are produced through knitting and weaving technologies. Among these technologies knitting is the most common manufacturing process for the production of spacer fabrics (Figure 2).

There are two types of fabrics : warp knitted spacer fabrics and weft knitted spacer fabrics. The first type is knitted on a rib raschel machine having two needle bars, while the second is knitted on a double jersey circular machine having a rotatable needle cylinder and needle dial [2].

3D spacer structures are much like a sandwich the constructions comprising two separate fabric webs which are joined together by spacer threads of varying rigidity.

Where the face and back of the fabric are knitted separately, at the same time, it is interconnected by monofilament yarns.

Three layers of this fabric is constructed at the same time, so the cost of laminating or combining is reduced.

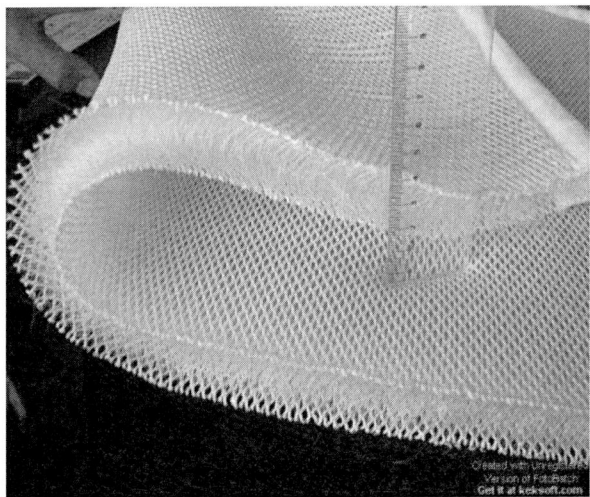

Figure 2 – Photograph of a knit spacer fabric [1]

1.2   Weft knitting

Ciruclar knitting machines with two sets of needles have the ability to create to individual layers of fabric that are held together by tucks. Such a fabric was referred to as a double faced fabric also be called as spacer fabric. It is produced by flat, v-bed and purl machines [2].

All techniques require the use of at least three different yarns for each course of visual fabric. The degree of space or height between the two fabric faces is determined in the circular knitting machine by the setting of the dial height relative to the machine cylinder. Spacer fabric heights preset in this way can vary between 1.1 and 5.5 mm.

With the uses of high and low butt needles, spacer fabrics are produced in weft knitting by tucking on dial and cylinder needles at the same feeder and knitting/plating on the dial needles.

At feeders 1 and 3 spacer yarn knitted on dial needle and tucked on cylinder needles.

1.3   Warp knitting

Warp knitted spacer fabrics consist of two ground surfaces, which are bound through pile yarns with stitches. The production in Raschel machine with two needle bars is possible and has great similarity to flat knitting [3].

Great flexibility is associated with warp knitted spacer fabrics because different materials may theoretically be used in guide bars 1,2 and 3 and 4,5, and 6.

Raschel spacer structure has more advantages as it can be produced with dimensional and/or stretch. Also the air and water permeability of the structure can be controlled [3].

In this method we can produce different width of spacer fabrics without ripping or ravelling structure.

1.4   Properties

The force required to press the two surfaces together(pressure resistance)is dependent on the mass of the mono filaments in the structure, which means yarn count, stitch density and machine guage in relation to the spacer fabric [4-5]. Some other properties of warp and weft knitted knitted fabrics are as given below

1.5   Physical properties

- Excellent compression elasticity
- Breathability/Air permeability
- Cushioning
- Insulation

- Good bending performance
- Drapability
- Adjustable vapour transport
- Recyclable/Latex free

Aesthetic and comfort

- Age resistant
- Sterilization capabilities
- Surface resistance
- Wash resistance
- Temperature regulation
- Light weight
- Diverse surface design capabilities

Other properties

- Elimination of cut and sew operations
- CAD usage
  Design flexibility
  Rapid prototyping
- Yarn diversity
  Increased attributes(i.e., wicking , extensibility)
- Variety of end capabilities

Application

- Automotive textile
- Home textile
- Sports clothing
- Sports clothing
- Medical textile
- Technical textile

The 3D speciality structured spacer fabric is a new emerging field for the new product developments, as their differential characteristics offer more extensive conditions for varied applications.

1.6   References

[1]   Shanna MB, Nancy P, Gary S, Three dimensionally knit spacer fabrics – A review of production techniques and applications, JTATM, Volume 4, Issue 4, Summer 2005.

[2]   Ajgaonkar DB, Knitting technology, Universal publishing corporation, 1998.

[3]   Warp knitting technology for production of technical textiles and their applications, ITB fabric forming, Volume 37, 1991.

[4]   Ulrike S, Use of warp knitted spacer textiles in mattresses, Melliand international, May 2006.

[5]   Ahmet U, Gerald H, and Chokri C, Development of weft knitted spacer fabrics for composite materials, Melliand international, May 2006.

Chapter 2

Comparison between various types of knit spacer fabrics

Summary

Spacer fabrics are 3D fabrics that comprise of two outer textile substrates that are joined together and kept apart by an insert of spacer yarns, mostly monofilaments. This creates a ventilated layer of air, allowing heat and moisture to escape. One reason for development of spacer fabrics was an attempt to replace toxic, laminated-layer foam with a single, synthetic fibre type fabric, thus facilitating future re-cycling. An important advantage is the low weight in proportion to the large volume. The application areas of spacer fabric are unlimited ranging from healthcare, safety, military, automotive, aviation and fashion. Currently it is being largely used for functional clothing comprising sports shoes, bra cups, shoulder pads, knee and elbow protectors etc. This chapter deals with the manufacturing and applications of spacer fabrics and the developments in this field.

2.1 Introduction

As the textile complex is faced with increasing competition, innovation and specialization have been employed by many machinery and product manufacturers to create a niche in the marketplace. In an effort to compete and appeal to the end-use market, products that go beyond the current range of performance and style have been developed [1]. Through continuous improvement of both the technology and the produced fabric quality, both warp and weft machines are now capable of producing spacer fabrics to the very highest standard of quality for a wide range of applications. Using new possibilities and improved machine technology, spacer fabrics have already become an established feature of many of the fabric collections presented by our international customer base. As far as technical textile is concerned, current tendency of the market to become saturated, and in the light of fierce international competition in the clothing, home and household textile markets, the textile industry is becoming increasingly interested in new technologies. Knitted spacer fabric represents a special product group within this sector because of their excellent crush resistance, breathability, and 3D appearance. [2]

2.2   What is Spacer Fabric?

Spacer structures are fabric constructions comprising two separate fabric webs, which are joined together by spacer threads or fibres of varying rigidity. The intermediate zone creates a layer of air, which has an insulating and thermoregulatory effect. Modifying the structure of the knitted construction can alter the amount of air incorporated in the assembly [3]. The yarn material used to join the plain ground fabric at the front side and the plain ground fabric at the backside with a defined "space" is mostly a stable, pressure tolerant material. Spacer fabrics are breathable, resilient, flexible, and soft. They come in a variety of weights, colors, textures, and designs.

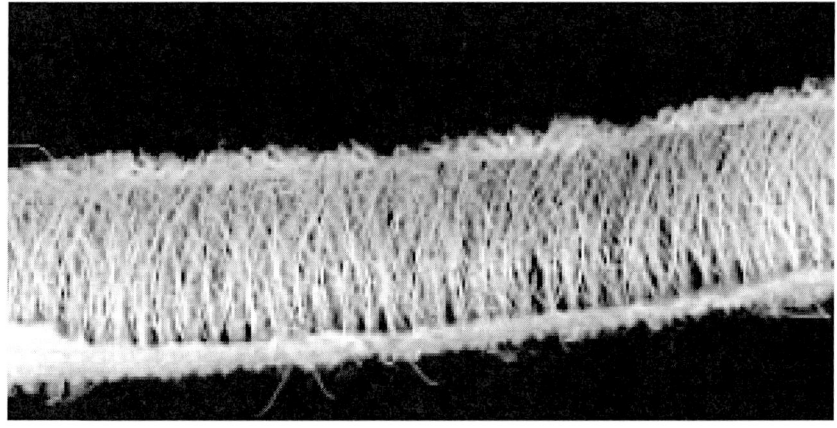

Figure 1 – Spacer fabric structure [1]

2.3   How Spacer Fabrics are different from Conventional Fabrics?

Spacer structures offer more extensive conditions for varied application, as their stable 3D structure encompasses scope for distinctive design and possess special physical attributes [4] (Figure 1), which are: • Lightweight, soft, pleasant on the skin • Offer active breathing properties • Pressure-elastic • Transport and absorb moisture (Figure 2) • Definable elasticity properties • Thermo-regulating properties • Wash-resistant • Ageing resistant and capable of sterilization

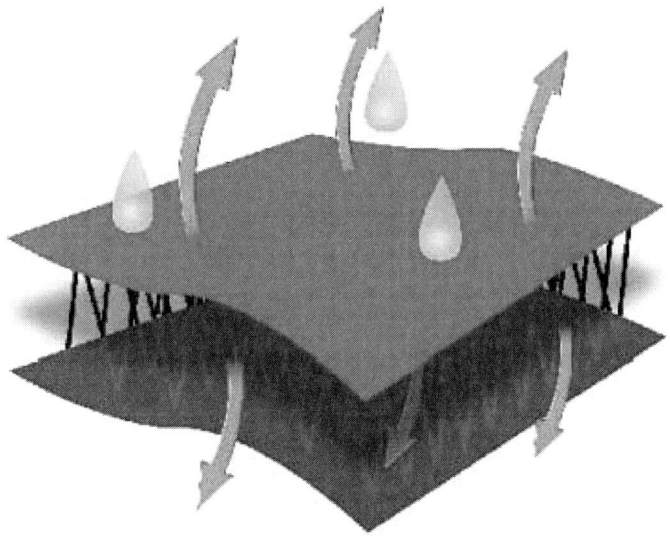

Figure 2 – Diagramatic representation of spacer fabric showing transport and absorption of moisture [3]

2.4   Knitting Techniques of Spacer Fabric Manufacturing

Followings are the knitting techniques used to manufacture spacer fabrics [3]- • Warp Knitting • Weft Knitting • Flat Bed knitting Through its unique manufacturing process, many different properties can be incorporated into the fabric to meet individual specifications [5] • Variable thickness • Variable compression • Rigid or Conformable construction • Varying degrees of open or closed knit • Waterproofing on one or both sides

2.5   Warp Knitted Spacer Fabrics

Warp knitted spacer fabrics consist of two separate fabrics linked by more or less stiff spacer yarns, generally monofilaments. The height of the space in the structure is determined by the set distance between the knock-over combs. The force with which the two fabric faces are kept apart is determined by the count of the spacer yarns, and also whether one guide bar is used or whether two guide bars are used in counter notation [6]. Knitting Construction [7] In warp knitting the needle bars are operating alternately and the guide bars swing between the needles of each needle bar in turn. Each guide bar shogged to make an overlap on each needle bar and so form knitted loops on the selected needle bar and a warp-knitted fabric is produced. Guide bar 1 is fully threaded and overlaps needle bar1. Guide bar 4 is also fully threaded and overlaps needle bar 2. Guide bars 2 and 3, which are threaded with spacer thread, overlap both needle bars. This way a warp knitted spacer fabric is formed.

Table 1 - Properties of Fibres used in Sacer Fabrics [1]

Fiber	Fiber type	Density (g/cm³)	Extension% (Dry/wet) (%)	Sp.Electric Resistance Ω/cm	Melting point, °C	Sorption (%)	Water holding	Water
Polyamide 6	Filaments	1.14	20-45/	10^9-10^{11}	215-220	3.5-4.5	10-15	105-125
Polyamide 6,6 Filaments	1.14	20-40/	10^9-10^{11}	225-260	3.5-4.5	10-15	105-125	
Polyester	Filaments	1.38	20-30/	10^{11}-10^{14}	250-260	0.2-0.5	3-5	100-105
Viscose	Filaments	1.52	10-30/	10^6-10^7	175-190	12-14	85-120	100-130
Cotton	Fibres	1.55	20-50	Low	----	7-18	42-53	

2.6   Knitting Construction [7]

In warp knitting the needle bars are operating alternately and the guide bars swing between the needles of each needle bar in turn. Each guide bar shogged to make an overlap on each needle bar and so form knitted loops on the selected needle bar and a warp-knitted fabric is produced. Guide bar 1 is fully threaded and overlaps needle bar1. Guide bar 4 is also fully threaded and overlaps needle bar 2. Guide bars 2 and 3, which are threaded with spacer thread, overlap both needle bars. This way a warp knitted spacer fabric is formed.

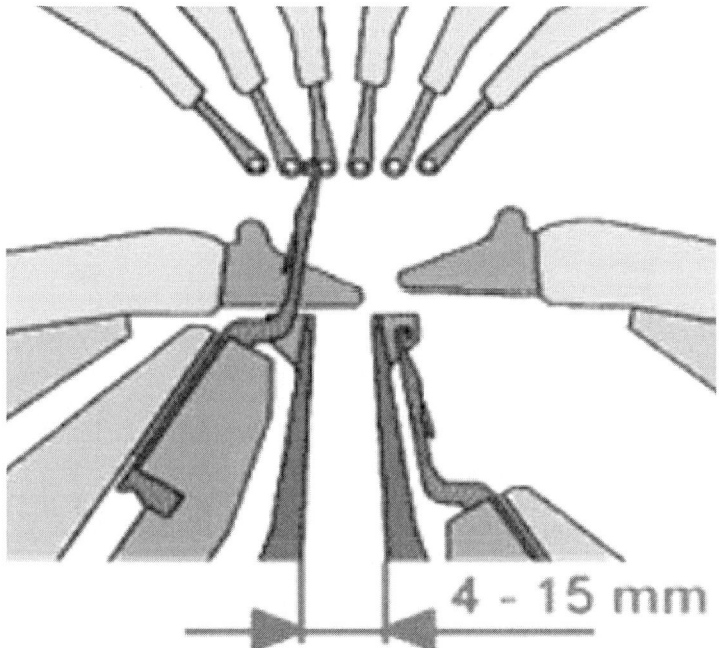

Figure 3 – Guide and needle bars in a knitting machine [2]

2.7   Features of Warp Knitted Spacer Fabric [8] •

Selectable spacing distance • Permanent elasticity • Durable elastic force • Good sound insulating properties due to porous surface • Saving in raw materials due to the entrapped air • Reasonable price because material consumption is lower • Easy to dispose of and environmentally friendly

2.8   Weft Knitted Spacer Fabrics [9, 10]

Spacer fabrics can be produced on circular double jersey machines as well as electronically controlled flat machines. The spacer threads are generally made of PES or PA monofilament yarns. The threads of the dial and cylinder stitches are made of textured PE filament yarns, depending on their intended field of application. The degree of space or height between the two fabric faces is determined in the circular knitting machine by the setting of the dial and height relative to the machine cylinder. Spacer fabrics preset in this way can vary between 1.5 and 5.5mm. The majority of spacer fabrics are currently produced using 8-lock circular knitting machines as it still offers the greatest scope for design versatility.

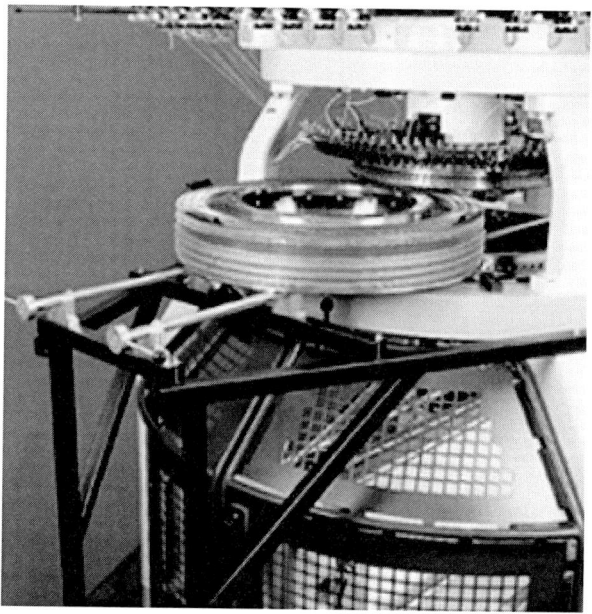

Figure 4 – Weft knitting machine [8]

2.9   Knitting Techniques (Figure 5 and 6)

Three different yarns are required to produce each course. These yarns are yarn for dial needles, yarn for cylinder needles and spacer yarn. Two different techniques for knitting these three yarns are described below: 1. Tucking on dial and cylinder needles at the same feeder (Fig 3) a. Tucking on dial and cylinder needles on feeders 1 and 4 (monofilament spacer yarn) on low and high butt needles alternately b. Knitting dial needles with dial yarn at feeders 2 and 5 on low and high butt needles alternately c. Knitting cylinder needles with cylinder yarn at feeders 3 and 6 on low and high butt needles alternately 2. Knitting/plating on the dial needles and knitting on cylinder needles (Fig 4) a. A special yarn feeder is required with two holes to enable two yarns to be knitted at the same feeder (feeders 1 and 3) b. Dial yarn knitted on the dial needles on low and high butt needles alternately; and spacer yarns knitted on dial needles and tucked on cylinder c. Cylinder yarn knitted on cylinder needles at feeders 2 and 4 on low and high butt needles alternately.

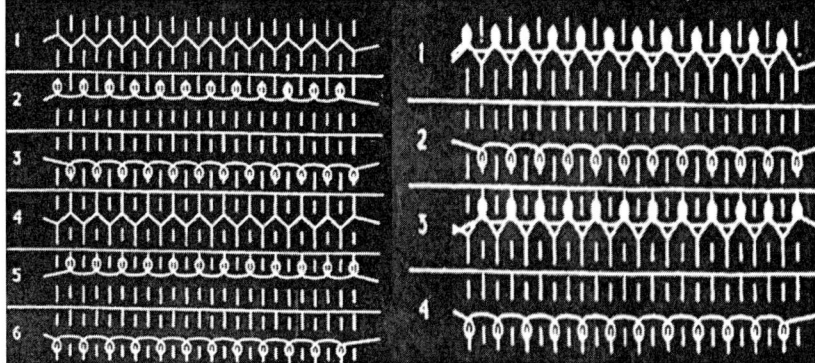

Figure 5. Tucking on dial and cylinder needles [3]   **Figure 6.** Knit/plating on dial and tuck on cylinder [3]

2.10   Advantages of Weft Knitted Spacer Fabric

• Plain as well as color and design and surface texture effects can be produced on the face of the fabric knitted by the cylinder needles • Various Shaped and true three-dimensional structures can be produced on electronically controlled flat machines.

2.11   Limitations

• The thickness of the spacer is normally limited to between 2 and 10mm • The basic structure of the spacer fabric is normally limited to either knitting the spacer threads on the dial and tucking on the cylinder, or tucking the spacer threads on the dial cylinder needles.

2.12   Warp versus Weft knitted Spacers [10]

i. Weft knitted spacer fabric is somewhat heavier, thinner and denser than the warp knitted fabric. ii. Warp knitted spacer fabrics show a higher thermal insulation in both dry and wet conditions. iii. The thermal absorption properties of warp knitted fabric are lower than the other fabric, which signifies that relatively the former is warmer to touch as compared to the weft knitted fabric, particularly when the fabrics have been wetted with water. iv. The water vapour permeability and resistance to evaporative heat loss properties of weft knitted fabrics are superior to warp knitted spacer material; therefore the former will be relatively more comfortable to wear next to the skin during a strenuous activity. v. Both spacer fabrics are more or less identical in their tenacity, breaking extension and initial modulus properties. vi. The absorption capacity and the wicking characteristics of the warp knitted spacer fabric are substantially superior to the weft knitted fabric. These may be due to two reasons:

a. Warp knitted spacer fabric is much more bulkier than the weft knitted spacer b. Textured polyester yarn was used in the warp knitted spacer fabric for the face and back fabrics whereas flat multifilament polyester yarn was employed in the weft knitted fabric for the two faces of the fabric. Monofilament was used in both cases as the spacer yarn.

2.13   Flat Knitted Spacer Fabrics

The Flat Knitting technology has been able to step out of its original areas of application. There is a growing interest in the technology, since the specific quantities of fabrics knitted on such machines as well as the exceptional way of production have been recognized for new technical particularly medical applications. Out of various forms of knitted fabrics, the V-bed arrangement allows production of spacer fabric, which is a tubular fabric construction with tuck stitch connections. The production of spacer fabrics is supported by positively controlled sinkers on both needle beds, which press down the last row of stitches during the formation of new stitches or other knitting operations [2].

2.14   Characteristics

• The thickness which can specifically be designed, depending on the gauge of the machine and the material selection (i.e. 4-4.5mm of thickness for a fabric coming from a machine of gauge E 14) • The different degrees of softness, which can be set in dependence of the angle of the inlaid tuck stitch and the chosen material. • The special fabric construction, which allows an excellent drapability; the fabric tends to keep the form, it is brought in. • The shaping ability similar to the regular fully-fashioned knitwear pieces. The spacer fabrics can also be knitted to shape.

2.15   Applications

• Cushioning elements for orthopedic use • Support fabrics in casting elements, which are used to support and fix broken limbs in defined positions. • Incontinence pads made out of superabsorbent fibres. • Special, ready to wear bandages with impregnated areas of curing emulsions.

2.16   Applications of Spacer Fabrics

3-D Automotive Textiles [11]

3-D textiles provide a perfect structure for car interiors because of softness, comfort, breathability, upholstery properties and aesthetic functions. They can be produced by two needled warp knitting, weft knitting, and flat knitting. Processing methods such as back spraying, back lining and laminating play an important role to impart required properties.

2.17 Car Seat Covers [12, 13]

For the seat comfort researchers investigated the possibilities of spacer fabrics for seat covering applications. The electronic jacquard machine was used for the development of the fabric. Initially, the research engineers developed fabric surfaces and designs similar to the appearance of fabrics used today for the manufacture of car seat covers. These fabrics have capability to completely replacing polyurethane foam lamination in the vehicle seat through the integration of PES monofilaments. The foam has drawbacks like flammability, lack of compression and resilience and delamination. The polyester materials performed.

2.18 Climatic Seat Comfort (Figure 7) [13]

The breathing activity and also the moisture buffering capacity of the car seat can be markedly improved if a warp knitted spacer fabric is used, particularly instead of the foam material. It's more open structure provides for the better conveyance of water vapour between body and seat, consequently producing better moisture management in the microclimate close to the body, which people immediately experience subjectively. Through the use of seat lining of a synthetic spacer fabric with an approximate thickness of 10mm, moisture management and heat conveyance are both optimized without forced ventilation by means of a fan.

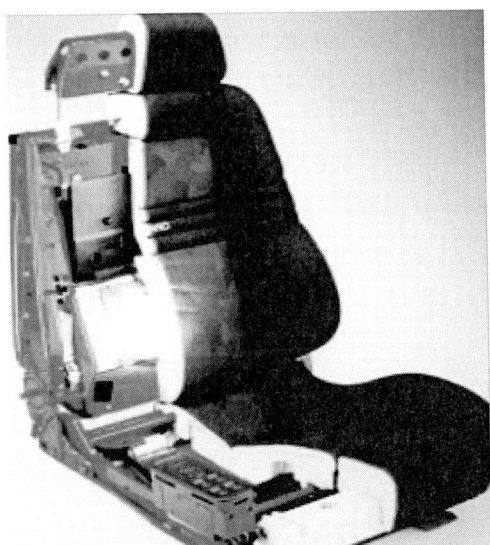

Figure 7- Knit spacer fabric for climatic seat comfort [9]

2.19   Spacer Fabric for Geotextile [14]

Knitted spacer fabrics offer an opportunity for Geotextile as their construction is exceptionally functional. By warp and weft knit spacer fabrics, it is possible to create the two outer layers with different structures such as a grid shape or mesh which is more effective at grabbing the soil than smooth fabrics. The spacer fabric offers a vertical filtration system for use in construction applications. The fabric can placed against a building and water then can drain into a sub grade.

2.20   Sports and Leisure application [15]

Sports and athletics have been increasing worldwide, as has for athletic apparel. In order to maximize the comfort and performance both moisture and temperature must be managed. This is possible with spacer fabrics. Spacers can offer different attributes like the www.fibre2fashion.com layer closest to the skin can be hydrophobic, the middle can be used for diffusion, and the outer layer can be hydrophilic thereby absorbing and evaporating heat energy. Spacer fabrics are also used in sports shoes because they are lightweight, high bulk, springy, washable, and a superior substitute for laminated foam. Gehring Textiles in St. Johnsville, NY makes a spacer fabric used in sliding shorts for baseball players.

2.21   Lingerie [16]

In the lingerie area too, spacer fabric has created a revolution of sorts and due to its comfort and moisture management/ wicking properties a number of companies are using it to create various lingerie collections, particularly brassieres. These brassieres are modified double knit structures in which both faces are joined during knitting process and require no further addition of foam or laminating aids.

2.22   Medical Application [17]

Properties such as heat and moisture regulation associated with both warp and weft knitted fabrics are ideal for use in medical textiles. An important field of application is wound care and prevention of chronic wounds, in particular pressure sores. There is a huge market for bandages and wound dressings. Spacer fabrics offer far superior characteristics to those of textile composites such as woven or nonwovens for medical applications such as orthopaedics, prevention of decubitus for the long -term bedridden, for surgical applications and convalescence, as well as for incontinence products. Other medical uses are Warp-knitted spacer fabrics for pressure relief, Multilayer compression stocking or bandage system, Bandage for the ankle joint, Absorbent diaper with therapeutic orthopaedic pad

2.23   Protective and Sports Clothing [18]

Other promising applications in prospect for spacer fabrics are opening up in the field of protective and sports clothing e.g. linings for firefighters' protective suits. The 3-D construction of the textile means that an insulating layer of air is formed between the two outer surfaces of the textile, which protects the wearer from the effects of heat, while at the same time guarantees that the garment is both breathless and comfortable. The flammability tests showed that the flame-resistant fabrics can be used both in protective clothing for firefighters as well as for mattress covers in the medical care sector, whereby their functional characteristics are not only determined by the use of flame-resistant materials alone. Their good compression behavior, moisture conductivity and thermoregulatory properties are also important criteria for guaranteeing physiological comfort.

2.24   Shoulder Pads [19]

Some analysts predict that these fabrics will replace neoprene in sports medicine applications such as shoulder pads, and knee/elbow protectors.

2.25   References

[1]   www.tx.ncsu.edu/jtatm/volume4issue4/vo4_issue4_abstracts.htm

[2]   M.Heide, "Spacer fabrics with specific protective characteristics", Melliand International, Vol. 6, June 2000, Pg 132-134.

[3]   "Spacers-At the Technical Frontier", Knitting International, July 2003, Pg 38-41. www.fibre2fashion.com

[4]   www.baltex.co.uk

[5]   www.heathcoat.co.uk

[6]   "RD 6N Raschel machine for spacer structures- product-oriented and highly productive", Kettenwirk-praxis, 4/94, Pg E3-E4.

[7]   www.karlmayer.de

[8]   Walter Schmidt, Krems, "The newly developed APL plush circular knitting machine", Knitting Technique, 14(1992) 3, Pg 163-169

[9]   "Warp knitted spacer fabrics- design and application fields", Knitting Technology, 4/2001, Pg 14-16

[10]   "Spacers-At the Technical Frontier", Knitting International, July 2003, Pg 38-41.

[11]   http://automotivetextiledesign.blogspot.com/2009/03/3d-spacer-fabrics.html

[12]   Dr RangaswamyVenkatraj, "Double needle bar warp knitted patterns", The Indian Textile Journal, March 1997, Pg 66-74

[13]   www.terrot.de/aktuelles/en/berichte

[14]   http://ara-texworld.com/index.php?view=article&catid=163%3Aind ustrialfabric&id=307%3Aknitting-fabrics-for-industrial-application- &tmpl=component&print=1&page=&option=com_content

[15]   http://www.inteletex.com/FeatureDetail.asp?PubId=&NewsId=191

[16]   www.faqs.org/patents/app/20080261490

[17]   Davies A et al, "The Use of Spacer Fabrics for Absorbent Medical Applications", Journal of Fiber Bioengineering and Informatics, Vol.1 No. 4 2009, pp321-330.

[18]   http://www.tx.ncsu.edu/ci/homeland/article_details.cfm?

[19]   http://www.human.cornell.edu/che/TXA/Outreach

Chapter 3

Stab resistance of warp knitted spacer fabrics

Summary

Warp-knitted spacer fabrics have been made out of UHMWPE fibers and subjected to quasi-static stab tests. The stab-resistant characteristic of the spacer fabrics and the effects of the fabric density and thickness on the stab resistance have been investigated, and the findings are as given below -

a) During the initial stage of stabbing, the stab-resistant law of warp-knitted spacer fabric is similar to the warp-knitted single-face fabric. The stab resistant behavior of warp knitted spacer fabrics invariably differed from other textile structures. Three kinds of deformations took place simultaneously during the process of resisting knife puncturing. These include tensile of surface knitted structure, shearing of yarns and compression of the spacer layer, of which the compressive deformation of the spacer layer proved crucial in stabbing.

b) The key parameters that influence the stab resistance are the thickness and density of warp-knitted spacer fabric. Even though there were differences in fabric thicknesses, the fabric density had the similar effect mechanism on the penetration force. The force of penetration can be increased and depth of penetration decreased with the increase in fabric density. There has been initial decrease in the penetration force followed by increase, with the increase in the thickness of warp-knitted spacer fabric. However, the fabric attains the highest stab resistance at a particular thickness. Through the compressive utilization of the compressive deformation of the spacer layer structure, the stab-resistant characteristic of the warp-knitted spacer fabric can be developed into the soft stab-resistant material. But, the compressive property of the spacer layer is measured merely from fabric thickness and density. Hence, further investigation is necessaryto determine the quantitative relationship between the compressive property of the spacer layer and the stab resistance. The design of the 3D structure of warp knitted spacer fabric is helpful to achieve the function of multiple protection.

3.1   Introduction

Stab-resistant armor is soft and chiefly comprises of woven, nonwoven or knitted fabric. Such textile structures possess various stab-resistant characteristics [1]. The good shear resistance of woven fabric has been attributed to high strength and high modulus fibers with dense structure [2]. The fabric in multi-layers is well resistant to sharp thorn, sticker, cone and so on. Owing to uniformity in yarn arrangement and high degree of freedom the unidirectional cloth made from aramid fibres exhibit a better ability in absorbing penetration [3]. But, it is necessary to combine with adhesive reagents so as to render an effective stab resistance under low-velocity stabbing. Non woven fabric has been found to possess a good shearing resistance and effectively absorb the penetration energy because of the entanglement of anisotropic fibers and the closed surface structure. However, as it is able to withstand less penetration force its used has to be restricted to only low-grade protection armor [4]. As the knitted has been easily penetrated by knife point and susceptible to take deformation while receiving puncturing, it has not been found to be so suitable. However, researches have also indicated that the knitted protective material possessed low weight, better designability, satisfying wide-area protection, and so on [5]. The multi-layer knitted fabric is able to absorb penetration energy, and has reasonably good shear resistance, wherein the stitches locked the knife to stop penetration before complete destruction of the fabric [6]. Owing to the deformation of weft loops and self-locking, the weft-knitted structure is able to resist stronger penetration force [7]. However, it was found that fabric had a larger deformation, and a deeper penetration. The structure and property of stab resistant warp-knitted single-face fabric has been studied [8]. The under loop structure has been unique to warp-knitted fabric and found to stabilize the stitch, and added to the accumulation of the yarns around knife edge, evidently proving beneficial in penetration force and yarn strength efficiency.

The aforementioned researches have shown that as the knife penetrated into the fabric, the textile structure underwent shearing and tensile action. Good stab resistance resulted from high-strength and good shearing-resistant fibers coupled with compact textile structure. On the other hand, the stab resistance can be improved by distortion of fabric which can absorb the penetration energy. The warp-knitted spacer fabric is a 3D structure which are stitch the upper and lower surfaces. The top and bottom layers are connected by monofilaments and form a bulky spacer layer [9]. The load has been transmitted to the spacer layer as the upper surface of the fabric was subjected to vertical pressure. The bending of the monofilaments absorbed the penetration energy resulting in compression of the spacer layer. For over years the spacer fabric having such features has attracted major attention in the area of cushioning [10]. Based on the potential of penetration energy absorption of the special spacer layer and the knitted-stitch structure, the study focused on the stab resistance of warp-knitted spacer fabric for the first time. Tests have been conducted to analyze the stab-resistant behavior of the spacer fabrics having various structural factors, and the findings pave the way for a new concept in the research and development of the soft stab-resistant materials.

3.2   Technical details

The fibres used for the soft stab-resistant body armor include ultra high molecular weight polyethylene (UHMWPE), Kevlar and poly-p-phenylene benzobis-thiazole (PBO). These possess properties such as high shearing endurance, dent resistance and high modulus characteristics. Under conditions where the predominant forces are tension and shearing, UHMWPE fibers have low density and outstanding performance under low-velocity stabbing [11]. Polyester monofilament has been used. As UHMWPE fibers are fluffy, high intensity and produce static electricity easily, they posed difficulties in production and hence have been given a protective under-twist of 50 twists/m, improving the cohesion in yarns without causing strength reduction. The yarns have been knit on raschel double needle-bar warp-knitting machine having six bars [16]. Sharkskin has been used for knitting the surface structure of the spacer fabric. Twelve fabrics have been knitted having various specifications. The yarns did not undergo sizing. Fabrics having various structures would behave differently during the stab-resistant process. The stab resistant property of the spacer fabric has been determined by quasi-static stab test [12]. The puncturing angle has been adjusted such that the knife was parallel to the courses to avoid the influence of the angle variation on the test results. During the testing process, the knife was dropped at a constant speed of 20 mm/min to penetrate the specimen. The end point of the test was reached as the penetration force attenuated to 90% of the initial penetration force. The knife speed has been controlled and also the curve of penetration force versus penetration depth, and penetration energy has been recorded. Observation has also been done on the deterioration of the textile structure.

3.3   Evaluation of stab resistantance properties

Based on the quasi-static stab test warp knit spacer fabric, the curves obtained relating to the penetration force versus penetration depth are depicted in figure 1. Four kinds of fabrics have been considered as follows

(a) Warp-knitted spacer fabric

(b) warp-knitted single-face fabric

(c) woven fabric

(d) non-woven fabric

The curve can be divided into three stages based on the puncturing process of the warp-knitted spacer fabric (Figure 1) during the penetration

During first stage, the pointed tip of the knife tip easily penetrates the upper surface of the fabric because of the easily deformed knitted stitch (Figure 1(a)). However, as the puncture opening in the fabric increased and stretched, the sliding yarns gradually gathered around the knife edge woven, warp knit single face and warp knit spacer fabrics [16].

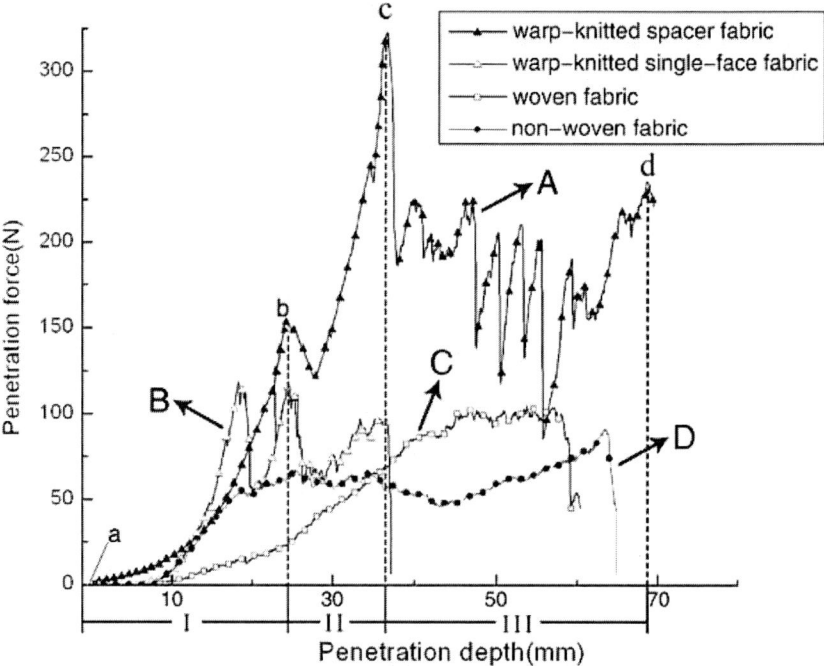

Figure 1 – Comparative curves depicting penetration force against displacement for woven, non

Hence, the fabric resistance to the knife has been significantly improved and the slope of the curve is larger during this stage (Figure 1). Simultaneously, the pressure received by the upper surface of fabric was transmitted to the spacer layer and the monofilaments near the knife edge bent to absorb the penetration of the knife edge. Hence, resistance of the fabric to the knife has been remarkably improved and the slope of the curve has been larger during this phase (Figure 1). Simultaneously, the top fabric layer received the knife pressure and transmitted it to the spacer layer and the penetration energy was absorbed by the bent monofilaments that were bent near the knife edge.

Hence, top layer of the fabric has been depressed [13]. After reaching a particular depth of penetration, the gathering yarns locked the knife completely and the penetration force attained its first peak point. The knife penetrates further by cutting the nearest constrained yarns.

During the second stage, the knifepoint got released due to some certain being cut, and there has been a decrease in the penetration. As the surrounding yarns broke one after another, the top fabric layer opened and expanded gradually and resulted in fluctuation of the penetration force. At the same time, the energy of penetration was absorbed by continuation of the spacer layer compression. The deformation of stitches and stretch of yarns increased the resistance to the knife as the knife point reached the bottom fabric layer. There has again been significant increase in the penetration force till the bottom fabric layer locked the knife for second time. The blade continued cutting the gathering yarns from the lower surface fabric. In the meantime, the top layer stitches acted as a wedge to the knife body. In this case, the penetration force reached its second peak point.

During the third stage, there has been compaction of the spacer layer, and the lower surface yarns around the knife got cut. An abrupt drop of the penetration force could be due to the simultaneous breakage of the top as well as the bottom surface yarns. But, the newly gathering unbroken yarns of both layers continued to offer strong resistance to the knife. So the curve A appeared a greater fluctuation in this phase until the upper and the lower surfaces were completely damaged and lost their effectiveness in the stab-resistance. Compared to the curves B, C and D of the single face fabrics, the curve A has been found that the curve A pertaining to the warp-knitted spacer fabric was more complicated (Figure 1). During the primary stage of knife penetration, the behavior of single-face warp-knitted fabric very much resembled to the warp-knitted spacer fabric during the first stage. The resistance to the knife increased with the deformation of stitches caused by the stretching and slipping of the yarns. There has been increase in the penetration force, which attained the peak point till the knife was tightly locked. Further, due to the broken yarns, the penetration force rose and fell. But, the spacer fabrics had mainly suffered the shearing effect of the knife during the whole puncturing process as depicted by the curves C and D pertaining to woven and non-woven fabric respectively. So the penetration force increased relatively gently and peak point was not obvious on both curves. Because of the special 3D structure, the warp-knitted spacer fabric showed the characteristic that has not been seen by the single-face fabrics and performed better during the stab resistance tests. The stab resistance of the knit spacer fabric has been influenced synergistically by the compression deformation of the spacer layer, the yarns stretching and shearing deformation of both surfaces and can change their roles during the entire process of puncturing. Finally, the energy of penetration has been effectively absorbed by compression of the spacer layer and increased the friction resistance to the penetration of knife as well.

3.4   Fabric density against stab resistance

Considering that the machine gauge is invariant, the density of take down is the determining factor in adjusting the warp-knitted fabric stitch density. Moreover, the tightness of the top and bottom layers determines the change of the take-down density and affects the density of monofilaments in the spacer layer. The compressive property of the spacer layer is subsequently influenced by the take-down density. Figure 2 depicts the relationships among penetration force, penetration depth, penetration energy and the density of take down. Irrespective of the distance of the knock over bar, there has been increase in the maximum penetration force and the penetration energy, and reduction in the penetration depth with the increase in density of take down.

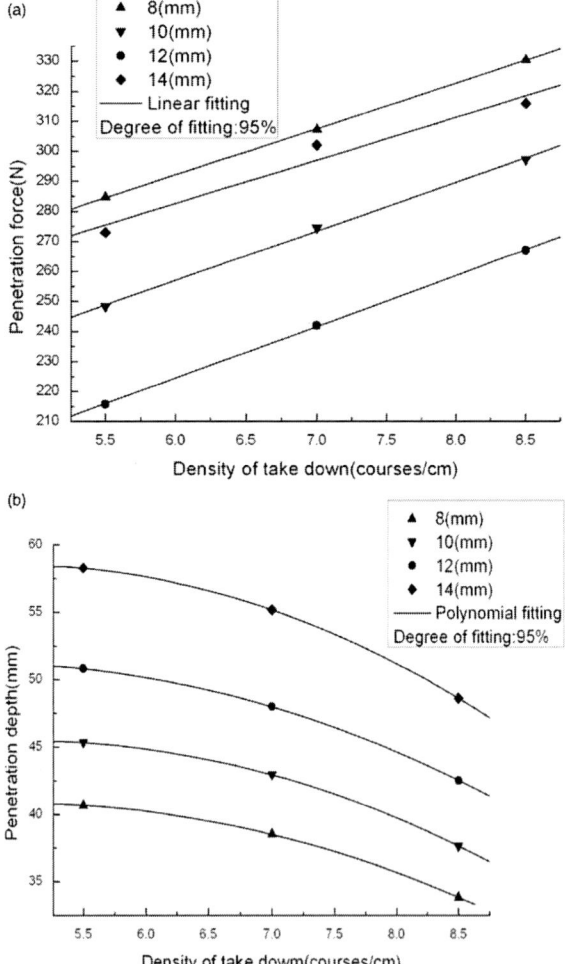

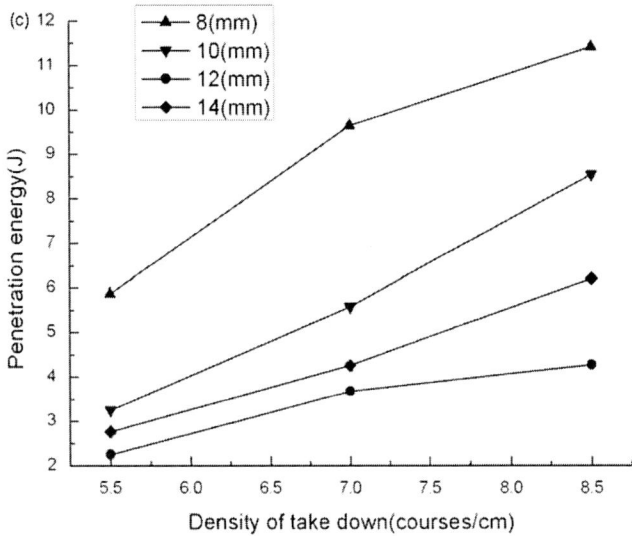

Figure 2 – The curves depicting penetration force against density of take down [16]
a)   The curve depicting penetration force against the density of take down
b)   The curve depictingthe depth of penetration against density of take down.
c)   The curve depicting penetration energy against the density of take down.

The increase in stab resistance has been due to increase of the density of take down, and also increased the stitch density of the upper and lower surfaces, and the tightness of cloth surface, accordingly. At the same time, the reduction of tensile deformability of the stitches and yarn slippage results in the yarns around the knife edge to be cut under minimal penetration depth [14]. The compression stiffness of the spacer layer has been improved due to the increase in the number of monofilaments loading. Hence, during the puncturing process the knife required more energy consumption. Considering the first and second stages (Figure 3), the high-density curve exhibited a higher slope with the increase in penetration force and higher peak values have been observed in the first and the second peak values. A larger curve slope indicated that the surface cloth was more difficult to be pierced. The penetration force value has been higher in this case. Hence, better performance has been observed in the stab resistance of the higher density warp-knitted spacer fabric. A linear relation has been found between penetration force and the density of take down. The line fitting method along with an equation has been used to determine the first-order linear relation of the penetration force and the density of take down. It has been found that the density of take down had similar effect on the penetration force, irrespective of the fabric thickness.

Also, the resistance offered to the knives by the different thickness of the warp knit spacer fabrics is reflected by variance of coefficient. With the increase in the take down density and the knock over bar distance, there in increase in fabric areal density. Under the condition that the thickness of warp knit spacer fabric remaining unchanged, as the areal density increases unevenly there is increase in the maximum penetration force.

$$\text{Specific force} = \frac{\text{Maximum penetration force}}{\text{Areal density}}$$

Since specific force is a key factor of the armor weight.

Figure 4 depicts the comparison of specific force of each test fabric so as to provide better rational evaluation to various structures. It can be observed that the specific force of fabrics are still higher. But, with the increase in the distance of knock over bar penetration force per weight has reduced. As the distance of knock-over bar increased, the penetration force of fabric increased obviously. However, the specific force has been less than that of fabrics of lower thickness. It can be attributed increase of fabric thickness which caused considerable increase in the content of polyester monofilaments. Hence, increase in the fabric weight lead to reduction in the penetration force divided by areal density. Also, since the density of take sown is 8.5 courses/cm, the penetration depths per weight are all lower. Under such situation, even with the same structure of same weight, warp-knitted spacer fabric would show greater stab resistance when the take down density is higher.

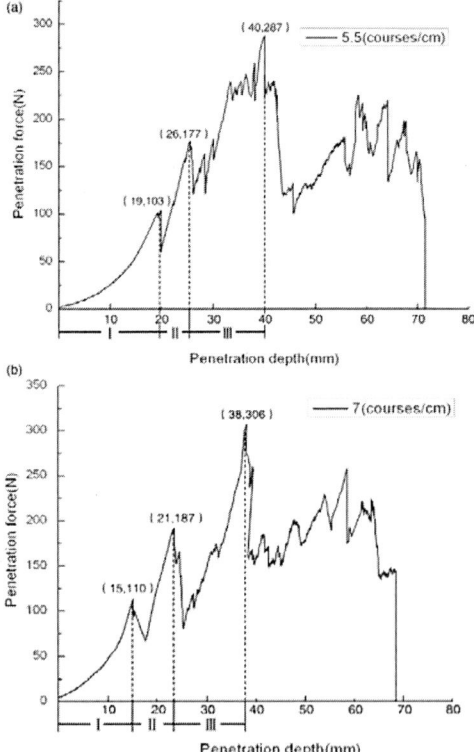

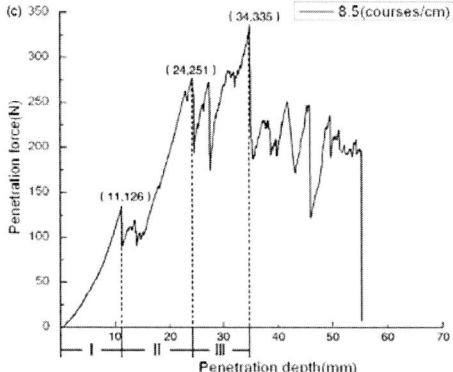

Figure 3 – Curves depicting stab resistance against density of take down [16]

a)   Penetration force curve against density of take down
b)   Penetration depth curve against take down density
c)   Penetration energy curve against take down density

Hence, the warp knitted spacer fabric having moderate thickness and higher density will exhibit best stab resistance when comprehensively considering the penetration force and depth per weight. Owing to the constraint relating to the model of machine, guage, and fibre fineness, the density of take down could only increase to a certain extent. Simultaneously, despite less fabric flexibility, the warp-knitted spacer fabric areal density as well as its production cost would increase along with the increase of the take-down density. Therefore, the density of take down should be

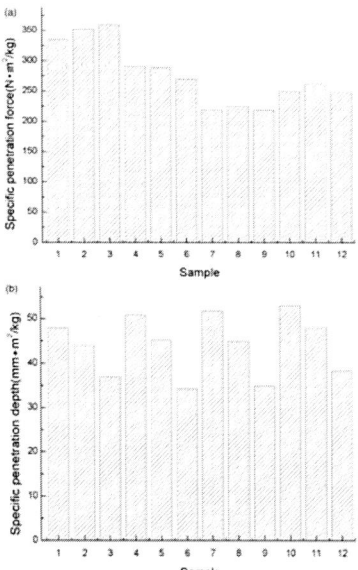

Figure 4 – Penetration force and depth/weight of various warp knit spacer fabrics

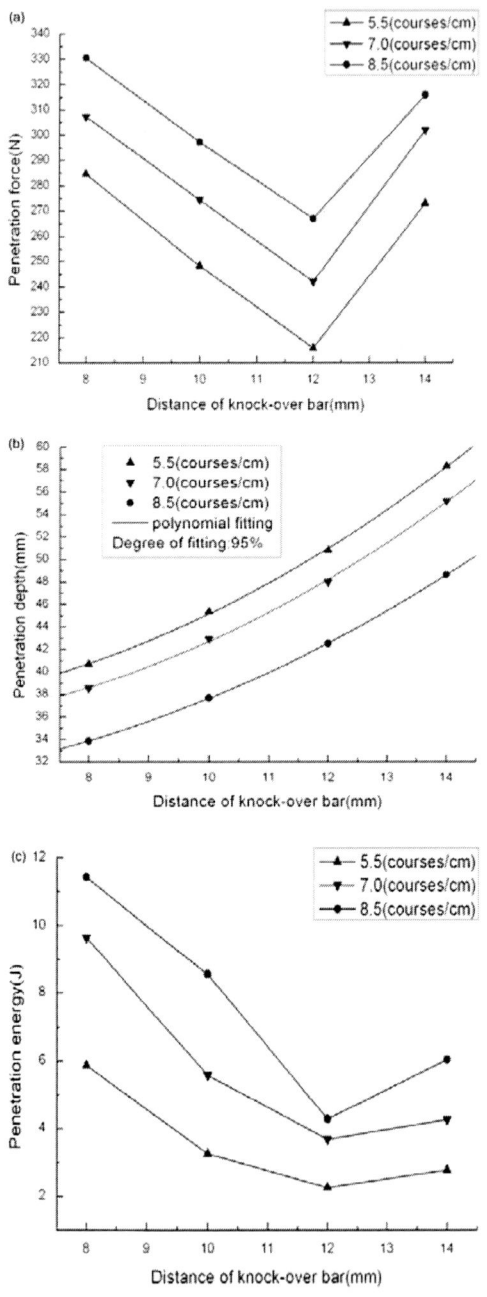

Figure 5 -Stab resistance curves against the distance of knock over bar [16]

a)   Penetration force curve against knock over bar distance

b)   Penetration depth curve against knock over bar distance

c)   Penetration energy curve against knock over bar distance

Hence, the density of take down should enable to strike a balance between the stab resistance of warp-knitted spacer fabric and its production cost and flexibility.

3.5   Spacer layer thickness and stab resistance

Investigation of the compressive property of warp-knitted spacer fabric shows that the spacer layer thickness is dependent on the distance of knock-over bar, and is considered crucial in affecting the fabric compressive property [15]. Figure 5 depicts the relation between penetration force, penetration depth, penetration energy and the distance of knock-over bar. The stab resistance of spacer fabrics has been investigated by selection of various thicknesses in the same density while ruling out the influence of the surface stitches on stab resistance [16]. With the increase in the distance of the knock over bar the penetration depth of different take-down densities show increase. But, the penetration force as well as penetration energy show initial decrease, and subsequent increase. The graph showing the penetration force versus penetration depth curves of four types of spacer fabrics with different knock-over bar distances is depicted in Figure 6.

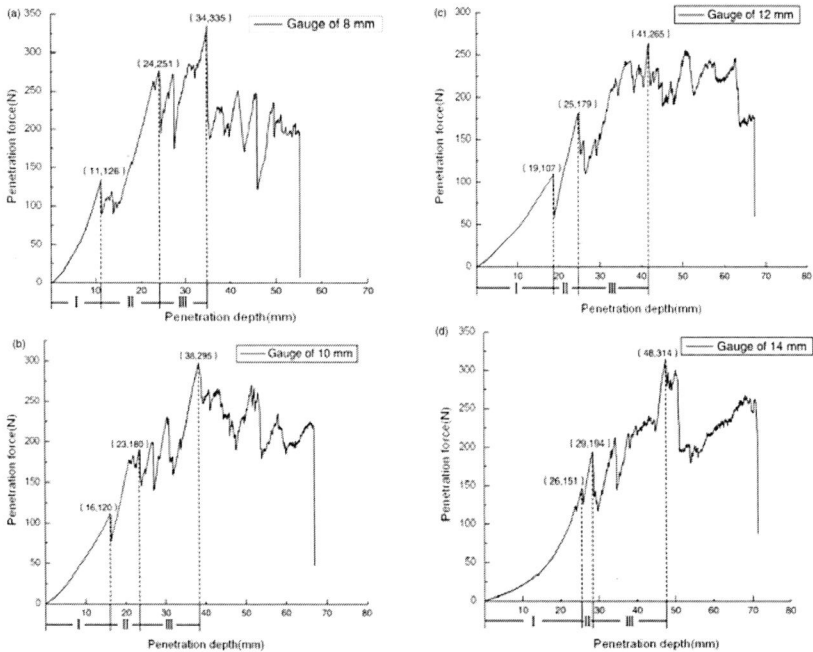

Figure 6 – Penetration force against displacement curves of test fabrics for 4 types of thicknesses in the same take down density [16]

   a)   Knock over bar distance -8mm
   b)   Knock over bar distance - 10mm
   c)   Knock over bar distance - 12mm
   d)   Knock over bar distance - 148mm

As can be seen from the figure during the phase I and II, with the increase in the penetration force the slope of the curve of higher knock-over bar distance was evidently lower. It suggests that the penetration of the knife into the spacer fabric has been made easy. Whereas during the puncturing process, the increase in the migration distance from the upper to the lower surface for the knife has been due to increase of the distance of knock-over bar. With the compression of the spacer layer, the spacer monofilaments could be bent easily. Hence, as the compression stiffness reduced the resistance of the spacer layer to the knife penetration decreased. Thus, by varying the distance of the knock over bar the penetration force and penetration energy reduced. However, as the distance of knock over bar increased beyond a certain limit the penetration force and energy increased. During the first phase the first peak point appearing at the moment that the knife was locked by the surface stitches (Figure 6(d)) has not been clear. It shows that prior to penetration of the knife into the stitches there has been easy compaction in the spacer layer. With the increase in the spacer layer thickness there has been reduction in compression stiffness. Hence, the top and bottom surfaces have been close to each other. The warp-knitted spacer fabric has been compressed to compact state. The resilience force of the spacer fabric was released obviously. This has resulted in more fibers taking part together in the puncturing process. Thus the penetration force and energy had the opposite trend to increase. But, with the appearance of the second peak point of the penetration force the penetration depth has been much higher (Figure 6(d)). By this time, the knife has penetrated into the fabric, which did not make much sense for stabbing. Lastly, there has been increase in the thickness of warp knit spacer fabric with decrease in the stab resistance under the given test specifications. It is due to the lesser influence of the compression of the spacer layer on resisting the penetration of the knife. Also, setbacks in the practical use have been caused due to the increase of the fabric thickness which also influenced the rise of fabric areal density and its volume.

3.6   References

[1]   Zhaowen GU. Study on the principle of soft complex stab-resistant body armor. J Text Res 2006; 27(8): 80–84.

[2]   Duong Tu Tien, Jong S Kim and You Huh. Stab-resistant property of the fabrics woven with the aramid/cotton core-spun yarns. Fiber Polymer 2010; 11(3): 500–506.

[3]   Fang Xinling and Zhang Yanpeng. Effect of resin on bullet-proof and stab-resistant properties of Kevlar UD cloth. Hi-Tech Fiber Applicat 2009; 34(3): 21–23.

[4]   Daozheng WU. The manufacture of soft complex stab-resistant material. Syn Fiber China 2011; 40(8): 32–34.

[5]   Gun Jinghe and Qiu Guanxiong. Research on the stab resistant performance of weft knitted complex stab-resistant fabrics, Master's Thesis, TianJin Polytechnic University, China, 2005.

[6]   Flambard X and Polo J. Stab resistance of multi-layers knitted structures (comparison between Para-Aramid and PBO Fibers). J Adv Mater 2004; 36(1): 30–35.

[7]   Yao Xiaolin, Qiu Guanxiong and Jiang Yaming. Research on the stab resistant mechanism of the weft knitted fabrics, PhD Thesis, TianJin Polytechnic University, China, 2006.

[8]   Li Lijuan, Jiang Gaoming and Miao and Xuhong. Structure and properties of stabresistant warp knitted fabric. J Text Res 2011; 32(4): 48–51.

[9]   Ye X, Fangueiro R, Hu H, et al. Application of warp-knitted spacer fabrics in car seats. J Text Inst 2007; 98(4): 337–344.

[10] Ye Xiaohua, HuHong and Feng Xunwei. Development of the warp knitted spacer fabrics for cushion applications. J Indus Text 2008; 37(3): 213–223.

[11] Xiong Jie, Shi Meiwu, Shen Tunian, et al. The progress of research on fiber assembly armor. Hi-Tech Fiber Appl 2001; 26(4): 11–16.

[12] NIJ Standard—0115.00, Stab Resistance of Personal Body Armor[S], 2000.

[13] Patrick MSchubel, Jyi-Jiin Luo and Isaac M Daniel. Impact and post impact behavior of composite sandwich panels. Compos Part A 2007; (38): 1051–1057.

[14] Tan KT, Watanabe N and Iwahori Y. Effect of stitch density and stitch thread thickness on low-velocity impact damage of stitched composite. Composites Part A 2010; (41): 1857–1868.

[15] Miao Xu Hong and Jiang Gaoming. The compression behaviour of warp knitted spacer fabric. Fibres Text Eastern Europe 2008; 16(1): 90–92.

[16] Miao X, Kong X and Jiang G, The experimental research on the stab resistance of warp-knitted spacer fabric, Journal of industrial textiles, 2012, 43(2)281

Chapter 4

Warp knitted fabrics for cushioning applications

Summary

Warp knitted spacer fabrics intended for cushioning applications have been produced on a machine having two needle beds and six yarn guide bars. Polyurethane foam (PU) has been chosen and compared with warp knitted spacer fabrics with regard to pressure distribution, air permeability, and heat resistance. The findings reveal that the warp knit spacer fabrics produced possess better properties than PU foam in respect to pressure relief. The spacer fabrics are more particularly suitable in warm conditions since they also possess very good air permeability and lower heat resistance, and thereby offer much better comfort than PU foam. Moreover, recycling is easier for the cushions produced with warp knit spacer fabrics. The investigation has shown that the warp knit spacer fabrics can replace PU foams for cushion application where there is much need for comfort and recycling.

4.1   Introduction

Owing to its impressive compression properties, PU foam has been well utilized as padding material over the years in applications such as seat, sofa, and mattress production. But, PU foam is not found to be such a suitable material in such applications. PU foam also poses difficulties with regard to recycling and comfort apart from presenting problem in washing [1,2]. In general the air permeability of PU foam is very poor. In certain situations, particularly in warm environment PU foam creates the feeling of being much warm and fuggy to the user. It also emits poisonous gases when burnt. Since PU foam needs isolation from other materials, disposal of its products poses a great problem, which renders its recycling process to become very cpmplex. A number of efforts have been directed towards overcoming these setbacks. One such example is the use of special cutting process to impart PU foam 'open' surface and by adapting high porosity to improve its 'breathability' [2]. Despite such efforts, most of the problems associated with PU foam cannot be completely solved. The use of technical textiles has shown a drastic growth during recent years. Certain conventional materials can be replaced by these fibrous materials that have many technical end uses and holds merits in particular end uses [3]. Warp knitted spacer fabrics (WK spacer fabrics) belong to this category and they are very interesting structures that can replace conventional PU foam used in seats, sofas and mattresses, and so on. The good compressive characteristics, air permeability, and thermoregulation by their unique 3D structure constitute the merits of WK spacer fabrics. WK spacer fabrics can easily be recycled because they are also made from fiber materials, and thus they can overcome the recycling problem presented by PU foam. Despite many studies on WK spacer fabrics, some elaborate studies related on the thick WK spacer fabrics for cushion applications have been reported. This chapter highlights the aim in developing this kind of WK spacer fabrics, which can replace PU foam for cushion application. The properties of PU foam has been compared and assessed with the developed spacer fabrics.

4.2   Structure, properties and knitting process

There are two surface layers in a WK spacer fabric and spacer yarns as the layer in between. A 3D structure results due to the spacer yarns connecting two surface layers (Figure 1). In the case of cushion application, the spacer yarns help to prevent the 3D structure from getting crushed due to pressure of the body. The choice of the spacer yarn having suitable bending rigidity and connecting method between two surfaces is obviously crucial in the design of the structure. A warp knitting machine having two needle beds and six yarn guide bard are used to produce the warp knitted spacer fabrics are normally produced on the warp knitting machine with two needle-beds and six yarn guide bars. As shown in Figure 2, while bars 1, 2 and bars 5, 6 are respectively used to knit the front and back surface layers, bars 3 and 4 are used to knit the spacer yarn layer.

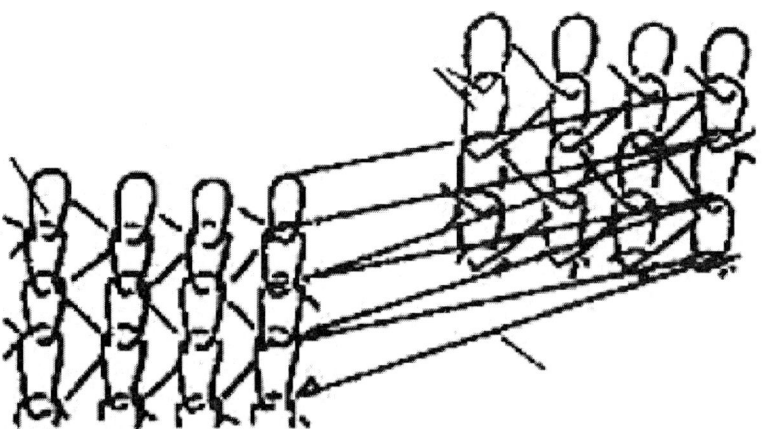

Figure 1 – Structure of a warp knit spacer fabric [7]

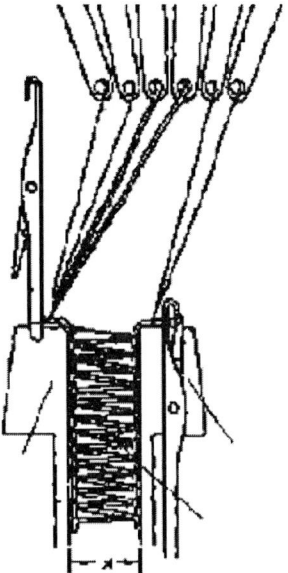

Figure 2 – Needle bars in knitting machine [7]

The two surface layers are connected together by feeding the spacer yarns through bars 3 and 4 on both front and back needles. They normally make symmetrical lapping movements to get better connecting effect. The two surface layers of the fabric can be connected by a number of methods. Two methods of connecting the surface layers are depicted in Figures 3 and 4. The vertical spacer yarns connect the two surface layers as depicted in figure 3. But, such a structure lacks stability as the spacer yarns tend to incline along the horizontal direction under the pressure. As depicted in figure 4 the structural stability can be improved by connecting the two surface layers with two systems of the symmetrical inclined spacer yarns. For fulfillment of varied end use applications, the change of the underlap amounts of the spacer yarns enables various inclination angles of the spacer yarns

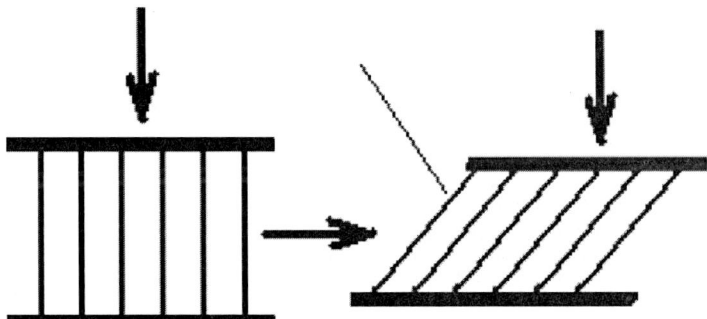

Figure 3 - Method of connecting the surface layers by spacer yarns vertically [7]

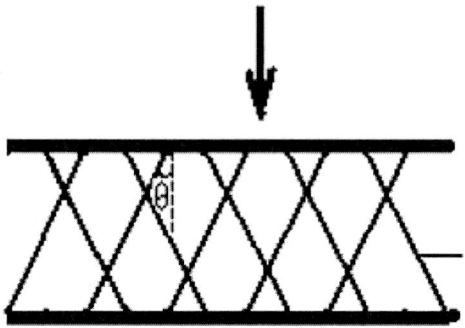

Figure 4 -Symmetrical inclined spacer yarns [7]

A bigger overlap obviously results in a higher inclination of the spacer yarns. Apart from various techniques of connecting available, the two surface layers of a spacer fabric can be knitted with same or different structures. Spacer fabrics are generally knitted with two surface layers having identical structures. The structures of the surface are either meshes or plain structures having various opening sizes. A mesh structure is one of the largely used structures that is formed with pillar and laying-in yarns (Figure 5). Such a structure is largely adopted in the spacer fabrics for cushion application since it provides good structural stability as well as good air permeability.

Another important structural characteristic of a warp knit spacer fabric is the space between two surface layers. Greater fabric thickness is required for cushion application. The thicknesses of cushions can range between 10-100mm or more based on various needs of application. But, the distance between the two needle-beds of the warp knitting machine restrict the thickness of WK spacer fabric.

Figure 5 – Knit mesh structure [7]

Presently warp knitting machines are able to produce spacer fabric upto thickness of 65 mm. In warp knitting machines distance between two needle beds can be varied upto certain limit. In order to achieve greater thickness of warp knit spacer fabric as in the case of certain cushion applications, 2 or more layers of WK spacer fabrics have to be combined since it is increasingly difficult to knit the spacer fabrics having very high thicknesses.

4.3   Development of the spacer fabrics

Raschel knitting machine having two needle beds and six guide bars have been used. A certain distance between needle beds has been fixed for producing the fabrics even though the distance could be increased [7]. PES multifilament has been chosen for surface layer yarn and two kinds of PES monofilaments for spacer yarns respectively. In order to knit the surface layers a specific linear density has been chosen based on the machine gauge and thickness of spacer yarn. But, yarns having high bending rigidity are required for the spacer yarns, since they have to support the body pressure.

Hence, PES monofilaments of two diameters have been chosen for the spacer yarns for knitting the fabrics having varsious compression characteristics. The surface structure shown in figure 5 has been chosen for all the fabrics. But, for connecting the two surface layers, three different underlap amounts of the spacer yarns have been used to achieve the different inclination angles. In the knitting machine one spacer yarn guide bar makes the lapping movements and the other spacer yarn guide bar makes the symmetrical lapping movements. Warp knit spacer fabrics have been produced by combination of different spacer yarn diameters and different underlap amounts. In order to increase the structural stability, the WK spacer fabrics after knitting are normally subjected to a heat-setting treatment. The heat-setting a thicker spacer fabric poses problems in a conventional heat setting machine as it is difficult to firmly hold the two selvedges of the spacer fabric. The selvedges of either sides of spacer fabrics have been knitted without spacer yarns so as to overcome this problem. The heat-setting temperature ranged between 160 to 190oC for duration of 1-2 minutes [4]. The spacer fabrics having longer underlaps of the spacer yarns or higher thickness or weight have been difficult to heat set since it was not easy to extend them to the desired width. The inclination angle of the spacer yarns is defined as shown in Figure 4. The thicknesses of all the fabrics are smaller than the distance set between the two needle-beds of the machine. It is so since during knitting the spacer yarns are under tensioned states. After knitting, the spacer yarns tended to relax into bended forms. The reduction in fabric thickness is due to the bending of the spacer yarns. The fabric thickness and mass density depend on the underlap amount of the spacer yarns. Greater thickness and mass density of the fabric result from the bigger underlap amount of the spacer yarns. It is due to the fact that there is less bending effect in the spacer yarns having higher underlap.

4.4   Assessment of properties

Tests relating to pressure distribution, air permeability and heat resistance have been carried out for assessment of the properties of the WK spacer fabrics. Comparison has been done with a PU foam of specified thickness [7]. The PU foam has some thickness difference compared to that of warp knit spacer fabrics. On the other hand, the areal density of the PU foam is only about 30–50% of the developed spacer fabrics.

4.5   Distrubution of pressure

Flexibility and softness are necessary for the materials intended for cushion applications. On the other hand, in the case of people sleeping on a bed or seated on a chair for long duration it is necessary to avoid the concentration of the body pressure. For example, a patient seated on a wheelchair always bears the pain caused by the pressure on the injured body. In the case of patients who are immobile in the same position for hours, bedsores or pressure sores are also caused by long-term downward pressure of the body on a support [5]. Hence, the assessment of the pressure distribution under the body pressure becomes crucial, particularly in the case where the relief of the high pressure is necessary [6]. The testing device used was pressure measurement system composed of a sensor mat and a data acquisition and analysis system. The dimensions of the sensor mat and the distribution of the sensors have been chosen specifically. Initial study has been conducted when a person of 50 kg stably seated on a massive wood chair placed with the sensor mat but without any fabrics. This has been followed by conducting studies with the placement of each fabric on the sensor mat. The seating supported the entire body mass during the test as the back, feet, and hands of the person did not touch floor or any other objects. For every test the pressure distribution of the buttocks has been determined and recorded by the system after the stabilization of seating. The tests have been performed under a standard atmospheric condition. Figure 6 depicts the findings of the average and peak pressures for all tested fabrics. Various pressed areas that have been measured can create various average pressures, since the average pressure has been calculated by dividing the weight with the pressed area measured. As the buttocks directly rested on the sensor mat the test without placement of the fabrics show the highest peak pressure and average pressure. In the case of all the spacer fabrics the peak pressures are lower in comparison with PU foam having greater thickness, despite the PU foam having smaller average pressure than all the spacer fabrics. It implies that the spacer fabrics developed exhibit better property for the decrease of the pressure concentration, and can be used in seat, sofa, and mattress as an alternative to PU foams for special applications in which the relief of the body pressure is very essential.

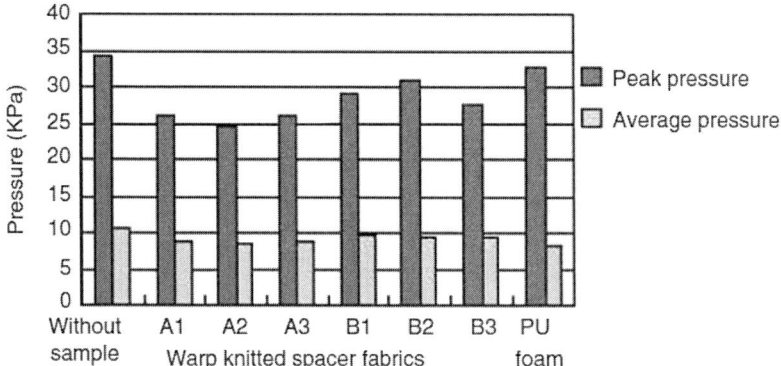

Figure 6 – Average and peak pressures for tested fabrics [7]

The findings also reveal that among all the spacer fabrics developed, the fabrics having with finer spacer yarns show lower peak pressure and average pressure values compared with the fabrics having thicker spacer yarns [7]. The differences between peak and average pressures have little significance in the case of the fabrics having the same spacer yarns, but with different inclination angles.

4.6   Air Permeability and Heat Resistance

In the case of cushion applications, air permeability and heat resistance are considered crucial factors since they considerably influence the comfort, particularly for some application cases, like car seats and medical mattresses. The air permeability test has been carried out based on the prescribed ISO standard. The heat resistance test has been conducted on special instrument under a standard atmospheric condition [7]. The testing results of air permeability and heat resistance have been determined. The findings reveal that the spacer fabrics exhibit very good air permeability. It is acceptable since the spacer fabrics have been produced with open surface structures. The air permeability for PU foam is far lesser compared with spacer fabrics, despite the PU foam being a little thicker than spacer fabrics. Low air permeability can cause a big comfort problem for PU foam used in special application cases as stated above. The tested PU foam has much higher value of heat resistance than the spacer fabrics. It implies that the PU foam has better capacity to retain the heat compared with spacer fabrics having same thickness. It can be an advantage for PU foams when they are used in a colder environment. But, the PU foam is rendered too warm and uncomfortable in warm conditions with a combination of lower air permeability and greater heat resistance.

4.7   References

[1]   Heide, M. (2001). Spacer Fabrics Trends, Kettenwirk-Praxis, 26(1): 17–20.

[2]   Prof. Dr. Umbach, K.,H. (2001). Physiological Comfort on Car Seats, Kettenwirk-Praxis, 26(1): 34–40.

[3]   Heide, M. (2000). Spacer Fabric with Specific Protective Characteristics, Melliand-Masche, 81(6): E124–125.

[4]   Anon (2004). New Machines for Heat-setting of Spacer Fabrics, Melliand- International, 10(4): 282.

[5]   Heide, M., Schurer, M., et al. (2002). Functional Warp-knitted Spacer Fabrics as Covers for Operating Tables in the Case of Long-term Operations, Kettenwirk- Praxis, 27(1): 25–27.

[6]   Ye, X., Hu, H. and Feng, X. (2005). An Experimental Investigation on the Properties of the Spacer Knitted Fabrics for Pressure Reduction, Research Journal of Textile and Apparel, 9(3): 52–57.

[7]   Xiaohua Y, Hong H and Xunwei F, Development of the Warp Knitted Spacer Fabrics for Cushion Applications, Journal of industrial textiles, 37(3), 2008, 213.

Chapter 5

Protective properties of warp knitted spacer fabrics under impact in hemispherical form

Summary

A warp-knitted spacer fabric intended for protection of human body against impact has been tested in the hemispherical form at various impact energy levels. Studies have been carried out on the impact resistance, energy absorption and force attenuation properties of fabric in the light of the experimental results. The various deformation and damage modes have also been studied by using frequency domain analysis coupled with Hilbert–Huang transform (HHT) technique. The following findings are summarized based on the experimental results and analysis:

a) The impact response of the spacer fabric is decided by the boundary condition.

b) The extent of impact energy determines the energy absorbed by the fabric.

c) In the case of a spacer fabric, its energy absorption capacity and the applied impact kinetic energy decide the peak contact force.

d) By the use HHT technique the deformation and damage modes of the fabric at various impact energy levels can be identified through the frequency domain analysis. As the impact energy increases, the frequency range distributed in the marginal spectra also increases.

The influence of fabric structural parameters and lamination on the protective properties of spacer fabrics under impact has been subsequently investigated. Warp knit spacer fabrics have been specially designed and produced for protective application and their transmitted forces have been measured. The influence of various structural factors like spacer monofilament inclination and fineness, fabric thickness, outer layer structure and fabric lamination on the peak transmitted forces have been explained and analyzed.

The protective performance of a spacer fabric is considerably influenced by its structural factors. The knit spacer fabric having greater inclination, coarser spacer monofilaments, greater thickness, and a more stable outer layer structure will have a better force attenuation capacity, if its destruction modes at different stages of impact energies are not different from others.

The influences of inclination of spacer monofilament and outer layer fabric structure on the force attenuation properties of the spacer fabrics can be altered by boundary condition in the hemispherical shape,

The force attenuation characteristics of spacer fabrics can be effectively enhanced by lamination. Three layers of the spacer fabrics knitted with a chain plus inlay structure for both outer layers in a total thickness of about 2.5 cm can comply with the European Standard BS EN 1621-1:1998.

5.1   Introduction

There has been a major focus on personal protective equipment during the past many years. They have been intended to protect wearers from different kinds of risks or hazards associated with their health and safety [1–7]. The energy absorbing material used as pads are the most popular among personal protective equipment, acting as impact protectors [8,9]. In order to prevent the human body from impact, blows or falls, these are integrated or inserted into protective clothing or equipment that are particularly designed for the purpose. In order to protect the various parts of the human body, many kinds of impact protectors are available. The protective performance of commercial impact protectors intended for different sporting applications have been assessed through development of series of standard tests so as to ensure required protection. As example, a number of European and British standards specify the requirements and test methods for hand, arm, leg, foot, instep, shin, chest, abdomen, genitals, trunk, head, breast and shoulder protectors for field hockey, martial arts, equestrianism, fencing, roller sports, football, motorcycling and cricket. The impact protection in such standards is normally measured by the force transmitted during impact. A striker having defined shape, size and weight drops freely onto the protector placed on a hemispherical anvil with the suitable velocity and impact energy. The zones of the body that require protection are simulated by the hemispherical anvils with various curvatures. The radius of the anvils lies between 12.5 to 150 mm, and the impact energies for the different kinds of protector range from 1 and 60 J. The most commonly used among European standards for motorcyclists impact protectors those for limbs and shoulders and those for back protection. As per the European regulations relating to personal protective equipment any clothing claiming to provide protection from injury should be tested and labeled such that it conforms to the relevant standard. During the recent times there has been a good deal of interest in the use of warp-knitted spacer fabrics in clothing and equipment providing protection against impact because of their combination of protection and comfort in use. Earlier investigations have concentrated on the static and dynamic compression behavior of a series of warp-knitted spacer fabrics [10,11]. Elaborate studies have been carried out on energy absorption performance and force attenuation capability of such fabrics under flatwise static and impact compression. Such investigations have shown that such fabrics possess the main characteristic of behaving as cushioning materials. There are three distinct phases involved in static and dynamic compression, which are linear elasticity, plateau and densification phases. However, in order to offer an adequate combination of protection and comfort, the protective material must conform to the shape and curvature of the body part being protected. Because of the alteration of boundary states during loading the impact properties of a protective material of curved shape obviously differ from those of a planar shape. Recently, an experimental study into the impact behavior of warp-knitted spacer fabrics of hemispherical shape has been reported [12]. The impact energy and weight of the striker have been maintained constant, and only the contact forces have been determined. The tests have been conducted as per prescribed European Standard. But, this standard specifies that protectors should be impacted using a striker of 5 kg weight at a kinetic energy of 50 J, and the transmitted forces then measured. Moreover, the investigation did not focus on the relation between energy absorption capacity and force attenuation properties, which is crucial to design fabrics to fulfill protective requirements. This chapter highlights the evaluation of the protective properties of warp-knitted spacer fabrics obtained from impact tests conducted strictly as per prescribed European Standard in hemispherical form, so as to simulate the realistic requirements of human body protection. The investigation has been done in two phases. In the first phase the evaluation of the impact behavior of a typical spacer fabric has been done so as to understand the impact process, energy absorption and force attenuation mechanism of the spacer fabric. In the second phase the influence of structural factors and lamination on the impact protective performance of the fabrics have been explained. The investigation could possibly enable a better understanding of the impact properties of warp-knitted spacer fabrics for protection of the human body.

Polymeric foams, rubbers, gels, and plastics are the materials generally used for assembling impact protectors [13–15]. But, such materials have serious setbacks. As a component of impact protectors, they exhibit poor comfort properties. Because of the lower air permeability and moisture transmission ability such materials cannot satisfy the comfort needs of majority of protective clothing used in sports and other extreme activities where sweat is easy to be generated and should be transmitted from the skin surface to the outer layer of the clothing, despite having a number of holes to improve their permeability. Also, such types of protectors have the constraint of being too heavy and possess low flexibility. Such materials are normally laminated by binding or gluing so as to get a better balance of protection and flexibility. But, the shock-absorbing capacity of impact protectors does pose a problem despite their wearing comfort and freedom of movement. Even though a number of impact protectors are commercially available in the market and their effectiveness in preventing injuries known, their use is limited owing to their poor wearing comfort. Warp-knitted spacer fabrics are consisting of two separate outer fabric layers joined together but kept apart by spacer yarns [16]. The striking structural characteristic is the space formed between the two independent layers of such fabric. Such type of knit spacer fabric serves well in impact protection owing to a combination of good out-of-plane flexibility, very good compressibility, high moisture conductivity, and good thermoregulation. The warp-knitted spacer fabrics can be an ideal class of energy absorbers for cushioning applications as pointed out by earlier investigations. Change of the structural factors enables their energy-absorbing capacity to be tailored according to specific end-use requirements [17]. But, there has been no report on systematic experimental study of the protective performance of warp-knitted spacer fabrics used as impact protectors in curved shapes for human body protection. Hence, in order to understand the manner in which the impact force attenuation properties of spacer fabrics in curved shapes are affected by the structural factors and lamination, it is necessary to conduct such an investigation. The impact behavior of a warp-knitted spacer fabric having a typical structure has been thoroughly analyzed [18]. The finding has revealed that the force attenuation of the spacer fabric is determined by its capacity to absorb energy and the relevant destruction modes under various kinetic energies, including plastic deformation, filament breakages, and thermoplastic transformation. Based on the evaluation of the frequency domain, such destruction modes relate to the specific frequency bands. The investigation has also shown that in comparison with the knit spacer fabric with planar shape the hemispherical shape has a lower efficiency force attenuation due to the spacer monofilaments of the fabric in the hemispherical shape which do not effectively resist the impact loading. Also, based on European standard for protection of limb, a single layer of the spacer fabric cannot offer adequate protection [19]. While preliminary investigation has been directed towards the energy absorption and force attenuation properties of a typical spacer fabric in a single layer, greater emphasis has been given to further investigate the influences of structural parameters and lamination of spacer fabrics. The investigation is supposed to provide useful information in the optimized structural design of warp-knitted spacer fabrics for human body protection against impact in curved forms.

Impact protectors are basically energy-absorbing material incorporated into protective clothing or equipment in order to protect the human body from impact strokes, blows, or falls [1]. Different types of impact protectors have been designed and used to protect various zones of human body like shoulders, elbows, hips, knees, and tibias from impact injuries in intense sports, like motorcycling, cycling, horse riding, skiing, skating, rugby, and hockey [1–3]. There has been an increasing demand for impact protectors due to their wide areas of applications. Materials like polymeric foams, rubbers, gels, and plastics are generally used for assembling impact protectors [1–3]. But, such materials have serious setbacks. They show poor comfort properties, when being considered as a component of impact protectors. In the case of majority of protective clothing that are used in sports and other extreme activities, and require easy generation and transmission of sweat from skin surface to the outer layer of clothing, the lower air permeability and moisture transmission capability of these materials cannot satisfy such needs, despite being always punctured with an array of holes to improve their permeability. Another problem associated with the protectors made of these materials, is that they are too heavy and possess low flexibility. Such materials are generally laminated in two or three layers through binding or gluing in order to obtain a better balance of protection and flexibility. However, the shock-absorbing capacity of impact protectors is opposed to their wearing comfort and freedom of movement. Despite the fact that many types of impact protectors are commercially available and most athletes are aware of their effectiveness in preventing injuries, the use of impact protectors has been unacceptable by the customers because of their poor wearing comfort. Warp-knitted spacer fabrics are consisting of two separate outer fabric layers joined together but kept apart by spacer yarns [4]. The crucial structural aspect of such fabric is the space formed between the two independent layers. This type of knit spacer fabric is rendered highly suited for impact protection by a combination of good out-of-plane flexibility, excellent compressibility, high moisture conductivity, and good thermoregulation capability. Earlier investigations have revealed that warp-knitted spacer fabrics can be an ideal class of energy absorbers for cushioning applications and their energy-absorbing capacity can be so designed to satisfy required end-use applications by alteration of their structural factors [4,5]. But, there is a need for a systematic analysis on the protective performance of warp-knitted spacer fabrics used as impact protectors in curved shapes for human body protection. Hence, it is necessary for carrying out this type of investigation so as to understand the manner in which the impact force attenuation properties are affected by the structural parameters and lamination of spacer fabrics in curved shapes. Previously, the impact behavior of a warp-knitted spacer fabric with a typical structure has been elaborately investigated [6]. The investigation has revealed that the force attenuation of the spacer fabric is determined by its energy absorption ability and the relevant destruction modes under various kinetic energies, including plastic deformation, filament breakages, and thermoplastic transformation. Based on the frequency domain analysis, such destruction modes are related to the specific frequency bands. The analysis has also shown that the spacer fabric in the hemispherical shape has a lower efficiency in force attenuation compared to the planar shape since the impact loading is not effectively resisted by the spacer monofilaments of the fabric in the hemispherical shape. Also, as per European standard for limb protection, a single layer of the spacer fabric cannot offer adequate protection [7]. While the previous investigation lays emphasis on the energy absorption and force attenuation properties of a typical spacer fabric in a single layer, subsequent investigation on the influences of structural parameters and lamination of spacer fabrics has been the major focus on the study. Such an analysis can provide useful information in the optimized structural design of warp-knitted spacer fabrics for human body protection against impact in curved forms.

5.2   Technical details

In order to study the impact behavior a typical warp-knitted spacer fabric has been used. Raschel warp knitting machine having six guide bars has been used to knit the fabric. During the knitting, the binding of the spacer fabric structure has been done by means of a polyester multifilament [20]. The spacer yarn used is a polyester monofilament to connect the two outer layers of the fabric. The details relating to the chain notation and the yarns used for each yarn guide bar have been considered. A digital thickness has been used to determine the fabric thickness. A specially designed drop weight impact tester has been used to carry out the impact tests based on certain prescribed European test standards. A striker strikes down upon test fabric situated on the anvil. In order to simulate the curvature of human shoulder, elbow, knee, forearm, tibia or hip, the anvil is made of polished steel of prescribed height and has hemispherical surface of certain radius. For measuring the transmitted force the hemispherical anvil has been mounted on a massive base through a load cell, with certain sensitivity in line with its sensitivity axis. Also the acceleration of the striker has been measured by accelerometers so as to get more information during the impact process. The acceleration readings obtained were used for deriving the displacement of the striker (or the compression depth of the fabric) and the contact forces between striker and fabric. Charge amplifiers have been used to amplify the voltage of the impact signals. Signals from the charge amplifiers have been recorded through two channels for the transmitted force, and for the acceleration. The transmitted force and the acceleration have been captured by a high-speed data capture card. Practically as areas of the human body can be subjected to different levels of energy impact in different sports or other activities, impact energies between 1-60 J are necessary for testing impact protectors based on mentioned the prescribed standards.

In order to evaluate the influences of structural factors on the impact behavior, a number of warp-knitted spacer fabrics have been produced on double bar raschel warp knitting machine and having various fabric thicknesses, outer layer structures, spacer monofilament diameters, and inclinations. In earlier investigations, these fabrics have been used for the static and dynamic compression tests in the planar form [2]. On the other hand polyester multifilaments have been used to bind the structure in the knitting process for the top outer layer and for the bottom outer layer. Polyester monofilaments have been used as spacer yarns to connect the two outer layers together. The outer layer of the knit fabrics have been produced with four different structures, i.e. locknit, chain plus inlay, small-size rhombic mesh, and large-size hexagonal mesh. The two outer layers have been bound together by 3 different spacer monofilament inclinations relating to underlapping one, two, or three needles between the front- and back-needle bars. The thickness of the knit spacer fabrics have been tested with a digital thickness tester having the standard deviation.

The spacer fabrics laminated as one, two, and three layers have been tested with specific kinetic energies according to the method indicated by the European Standard for motorcyclists protective clothing against mechanical impact [19]. The testing device and conditions have been maintained the same in previous case. In order to achieve accurate results, specified number of samples has been tested for each type of spacer fabric and method of lamination. As the transmitted force is required by the standard to assess the impact force attenuation properties of an impact protector, only the results of transmitted forces are adopted for discussion of the effects of structural parameters and laminated layers. All the transmitted force–time curves presented are the most representative curves, and all the peak transmitted forces presented are the mean values of ten test results with standard error.

5.3   Study of impact force

Figure 1(a) and (b) correspond to typical acceleration and transmitted force signals for various impact energy levels. During the first stage the acceleration shows a lower value over a relatively longer duration, and in the second stage rises rapidly to a peak.

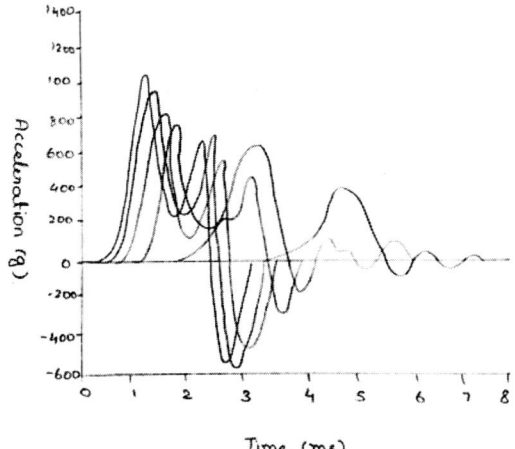

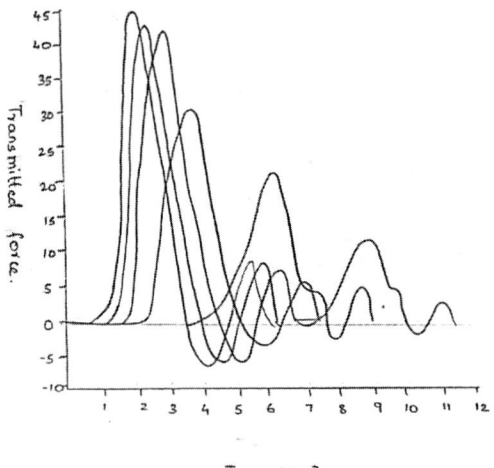

Figure 1 – Impact signals at various levels of energy [20]
a) Acceleration
b) Transmitted force

A higher peak in acceleration response is achieved over a shorter duration under greater energy. As per the momentum theorem, a shorter time of impact results in a greater acceleration and a greater reactive force. A very high acceleration and dynamic contact force is created due to the dropping striker being instantaneously stopped by the anvil in the absence of fabric. But, the striker undergoes deceleration as it strikes the fabric with the fabric located on the anvil surface. Initially the kinetic energy of the impact is stored and dissipated during the deceleration. It then releases the stored energy over a longer time, and thus results in a decrease in the acceleration of the striker and creates a smaller contact force between the striker face and the upper surface of the fabric. As the impact energy increases, the velocity of the striker increases during the initial part of its contact with the fabric. The time required to compress the fabric by the striker to its denser state is reduced. The fabric in its densified state is able to stop the striker more quickly owing to its higher compression resistance. Hence, the increase in impact energy at peak point will thus lead to a decrease in impact duration and an increase in acceleration. As depicted in figure 4(b), the force is transmitted to the anvil through the deformed fabric during the impact process and the transmitted force curves thus have a similar form to the acceleration curves. For a similar reason a higher level of impact energy will result in a greater transmitted peak force over a shorter duration. Figure 2 depicts the contact force–displacement curves obtained by integrating the respective acceleration curves. The magnified view of the curves is shown in the inset. There are two clear stages that can be seen in the contact force, just as in case of the acceleration and transmitted force signals. The first stage is visualized by very slow increase and the second stage is visualized by rapid increase. When the range of displacement is below a certain value, it is the stage wherein there is slow increase of the curves, relating to a compression strain of certain percentage. Further, there is a rapid increase in the contact force with the increase in the displacement. The curves are sensitive to strain rate, as indicated in figure 2.

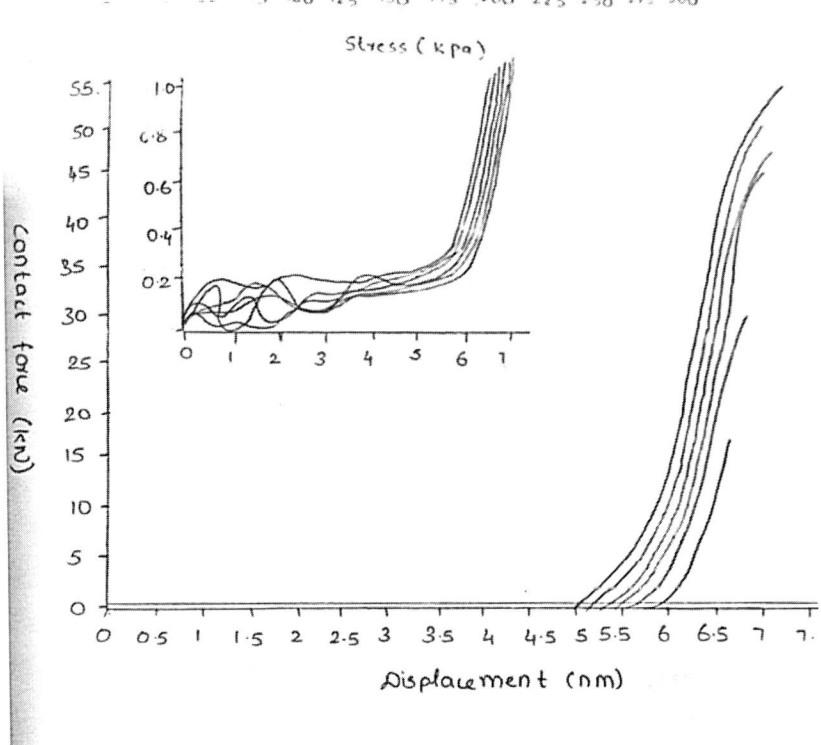

Figure 2 – Curves depicting contact-force displacement at various energy levels [20].

The change of impact energy indicates a change in initial contact velocity or strain rate since the height of drop of the striker controls the impact energy. The impact properties of the spacer fabric are also this sensitive to strain rate, since it comprises of polyester fibers. The fabric can be compressed more quickly into its densification stage and with a larger final displacement by a striker of greater kinetic energy, as depicted in figure 3. This is one of the reasons why the fabric has a higher peak contact force when it is impacted at higher kinetic energy. The impact behavior of the same fabric impacted in different shapes has been compared. Earlier investigation has revealed that there are 3 main stages involved in the force–displacement curve of the fabric under flatwise impact.

These are linear elasticity, plateau and densification stages [11]. But, in the case of impact in the hemispherical form the impact contact force–displacement curves reveal only two stages. There is absence of the plateau stage. Such difference arises from the various boundary conditions of the spacer fabric in planar and hemispherical shapes. The contact area between the striker face and the fabric surface remains constant during the complete impact process when considering impact in flat form. On the other hand, in the case of impact in hemispherical form, the contact area for the fabric is changed during the impact process and increases with the striker displacement. It can be confirmed through geometrical analysis.

A relationship can be derived from the geometrical analysis (Figure 4) by considering a cross-section of the fabric through the central axis of the anvil which is used to derive the contact area between the striker face and the fabric surface during the impact, based on the axis-symmetrical property of the fabric shape and the anvil. The following parameters have been considered in deriving the relationship

a)   Contact area
b)   Fabric thickness
c)   Anvil radius
d)   Displacement of the striker

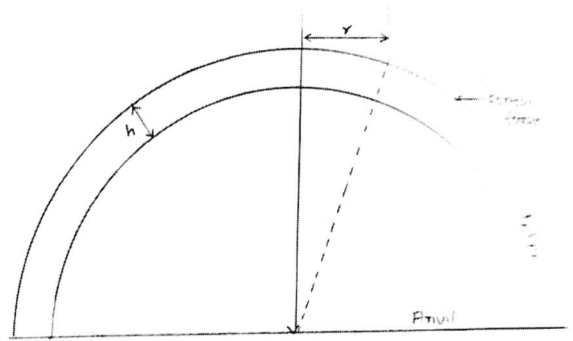

Figure 4 – Fabric cross section on the anvil [20].

From the values of the parameters, the variation in contact area with the displacement of the striker has been plotted (Figure 5). The contact area of the fabric with the striker rapidly increases due to the increase in the displacement of the striker. The contact force remains constant with the increase in the contact area. Hence, when considering the hemispherical shaped impact the plateau stage of the contact force–displacement curves obtained under the flatwise impact condition dies not occur.

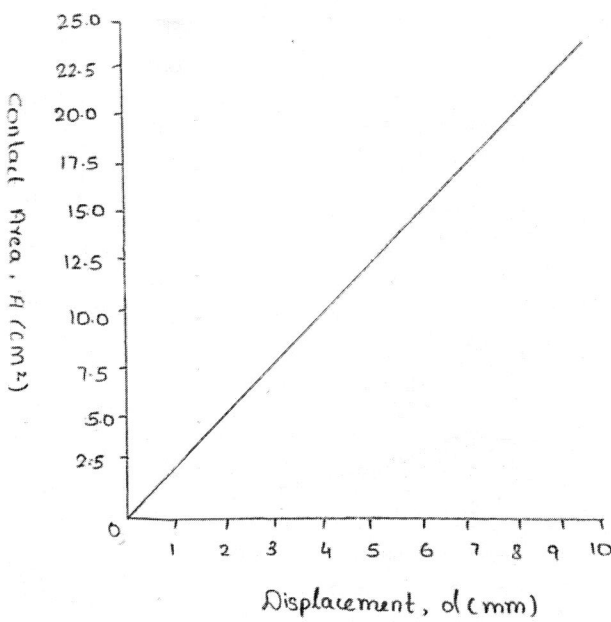

Figure 5 – Variation in contact area with striker displacement [20].

Also, the plateau stage is not being seen in the case of hemispherical impact since the spacer monofilaments within the fabric deform in different ways under flatwise impact and under impact of the fabric in the hemispherical form. Under flatwise impact all the spacer monofilaments simultaneously withdraw the impact load and are subjected to the same deformation. Whereas, the spacer monofilaments undergo various deformations at various positions, when considering the hemispherical form (Figure 6). In the case of flatwise impact, as the displacement of the striker is greater than the densification displacement of the fabric achieved, while in the case of the fabric impact in hemispherical form the linear elasticity, plateau and densification of the spacer monofilaments within the fabric can all be seen simultaneously. The deformation of the spacer monofilaments is greater as they get closer to the central axis of the anvil.

For a given displacement of the striker the distribution of strain of a spacer monofilament with the distance between two defined points for a given displacement of the striker. The strain of spacer monofilament has been found to change with distance between the two defined points for a given displacement of the striker in a parabolic manner. The maximal deformation is located at the central point of the contact surface. The distribution of strain of spacer monofilament with the distance between two defined points for a given displacement of striker for different values given displacement of striker in a cross-section of the fabric through the central axis of the anvil have been determined. As the displacement of the striker increases all the deformations on the contact surface evidently increase. In the case of impact in hemispherical form, as the spacer monofilaments within the fabric can deform differently, the plateau stage cannot be noticed in the contact force–displacement curves of the fabric.

5.4   Absorption of energy and force attenuation properties

When considering the hemispherical form, the impact process of the spacer fabric is highly complex. The kinetic energy of the striker at the time of the impact changes into the following forms of energy

a)   Energy absorbed due to deformation and damage caused to the fabric

b)   Energy absorbed due to the deformation of the type of testing (striker, anvil and supporting mechanism), and

c)   Loss of energy due to the stress wave propagation.

The total kinetic energy of the striker gets fully changed into deformation and dissipated energies during the point of maximal deflection and acceleration. This comprises of the following energies

a)   Energy stored due to the elastic deformation of the fabric and the type of testing

b)   Energy dissipated due to the plastic and damage to the fabric and the testing system

c)   Energy dissipated by the stress wave propagation.

Beyond this point, the striker will rebound due to the transfer of the stored energy to the kinetic energy. Ignoring the energy dissipated by the plastic deformation and damage to the testing system, and the loss of energy becayse of the stress wave propagation, a number of energy expressions can be obtained by considering the following energies

a)   Kinetic energy of the striker

b)   Energy stored by elastic deformation of the fabric

c)   Energy absorbed by elastoplastic deformation of the fabric

d)   Energy dissipated by damage to the filaments

e)   Energy absorbed by elastic and elastoplastic deformation of the fabric (Figure 5 - the area under the contact force– displacement curve)

f) Energy stored by elastic deformation of the testing system, and

g) Residual kinetic energy of the striker after being absorbed by the elastic and elastoplastic deformations of the fabric that has to be dissipated by the fabric damage or transferred to the testing machine..

The energy absorbed by the elastic and elastoplastic deformations of the fabric vs the striker displacement at various impact energy levels is depicted in figure 6. The energy absorbed is found to increase with impact energy. There is greater energy absorption by the fabric owing to its higher deformation, as impact by the striker at a greater degree of kinetic energy can lead to a greater final fabric displacement. The energy absorbed by the elastic as well as the elasto plastic deformation of the fabric and the residual energy of the striker, is depicted in figure 7. As the impact energy gets lower than a certain value, the kinetic energy of the striker is almost entirely absorbed by the elastic and elastoplastic deformation of the fabric causing no generation of heat. But, as the impact energy reaches beyond a certain level, the kinetic energy of the striker cannot be entirely absorbed by the elastic and elastoplastic deformation. This is so despite the energy absorbed by these fabric deformations of the fabric increases as the impact energy increases. Under such situation, the kinetic energy of the striker not absorbed by elastic and elastoplastic deformation of the fabric and is transferred to the residual energy. With the increase in impact energy, there is also increase in the residual energy (Figure 7). When the purely elastic and elastoplastic deformations of the fabric cannot totally absorb the higher kinetic energy of the striker, damage to the filaments takes place. It should be noted that while one part of the residual energy is transformed into heat, the other part together with kinetic energy of elastic and the elastic deformation energy of kinetic energy of elastoplastic makes the striker rebound. The fabric damage and deformation after various impact energy levels is depicted in Figure 8. At lower levels of impact energies, major plastic deformation of the monofilaments in the spacer layer takes place. The monofilaments get kinked, distorted or squashed under such situation. At a certain level of impact energy some breakage of the monofilaments takes place. But, in the outer layers of fabric there is no obvious deformation of the multifilaments. There is a certain extent of breakage of the multifilaments in the outer layers as the impact energy is increased to a certain value, besides the plastic deformation and breakage of the monofilaments. A hole in the outer upper layer of the fabric is created due to the impact, as the impact energy attains a certain value. Simultaneously there is breakage of a number of mono- and multifilaments. The mono as well as the multifilaments break and get fused together due to generation of heat, as the impact energy attains a certain value.

As the impact energy attains a higher value, there is breakage of the mono- as well as multifilaments and are rendered a rigid plastic by means of thermoplastic process. From the above discussion it can be seen that when the fabric is subjected to various impact energies it is deformed and damaged in various modes. The fabric entirely absorbs the impact energy as the energy absorption capacity of the fabric because of the elastic and elastoplastic deformations is greater than the kinetic energy of the striker. Under such situation, the stress-strain nature of the spacer fabric determines the contact law. Whereas, damage occurs to the fabric as it absorbs the balance kinetic energy of the striker as the impact energy is greater than the energy absorption capacity owing to the elastic and elastoplastic deformations. Under such condition, a very high contact force can be created since the contact law between the striker face and the fabric may change significantly, since the striker may collide directly with the anvil.

There have been no direct collisions between the striker and anvil under the investigation considered, since perforated holes have been absent in the fabrics after impact (Figure 8). Hence the contact stiffness of the compressed fabric at the rebounding point determines the peak contact force. In the case of that a highly densified fabric there is a high modulus and thus a high contact stiffness. Hence, the kinetic energy of the striker is preferably absorbed before the fabric is compressed to its high densification stage. With regard to earlier discussions, considering flatwise impact a longer plateau stage results [11]. Such response under elastoplastic deformation permits the fabric to absorb a great deal of kinetic energy at a lower constant contact force. Thus, when flatwise impact is considered the spacer fabric possesses superior force attenuation performance (figure 5). But, considering impact in the hemispherical form, the contact force curves increase in a monotonal manner with the displacement, since the spacer monofilaments within the spacer fabric in the hemispherical shape cannot effectively resist the impact load.

As depicted in figure 9, the graph plotted as energy absorbed by the spacer fabric vs contact force enables to understand the relationship between the force attenuation and the energy absorption behavior of the spacer fabric under impact in the hemispherical form. It shows that the energy absorbed increases non-linearly with the contact force. Hence it is necessary that the spacer fabric should be compressed with a higher displacement into a highly densified stage so as to absorb more kinetic energy. It results in higher contact stiffness, bigger area of contact, and a higher peak contact force between the face of the striker and the surface of fabric. Since all the spacer monofilaments are able to contribute to the resistance to impact loading, the force attenuation of the spacer fabric in the hemispherical form has lower efficiency than the impact in the planar shape.

The force of contact gets transmitted to the anvil by means of the deformed specimen as stress waves, under the impact. The process of transmission is complex and its mechanism out of scope under the investigation considered. The variation in the peak contact force as well as the peak transmitted force with impact energy is depicted in figure 10. The findings reveal that with the increase in the impact energy there is non-linear increase in both forces increase. Also, in the case of all impacts the peak transmitted force remains below the peak contact force. But, with the change in impact energy the ratio between them is not constant. The damping characteristic and stress wave propagation of the fabric is affected by the difference that can arise from their different modes of deformation and damage. As for example, in the case of the impact with a specific value of kinetic energy, the transmitting ratio is greater than those for other impacts.

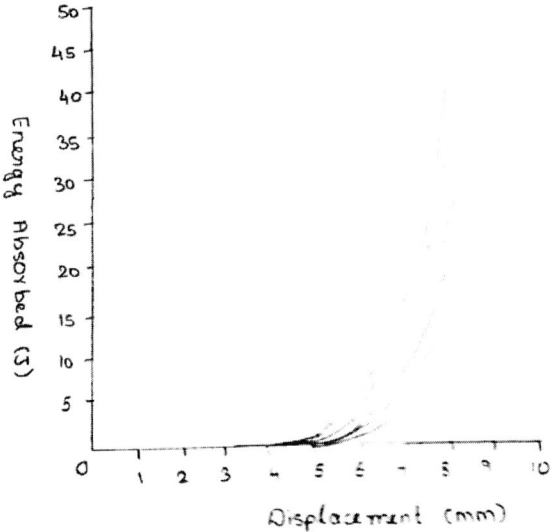

Figure 6 – Energy absorbed by the elastic and elastoplastic deformations of the fabric vs the striker displacement under various impact energies [20]

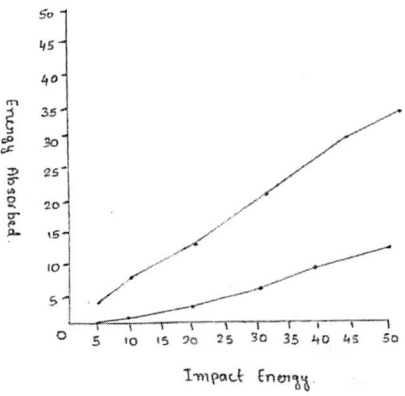

Figure 7 – Energy absorbed by the elastic and elastoplastic deformations of the fabric and residual energy for various impact energy levels [20]

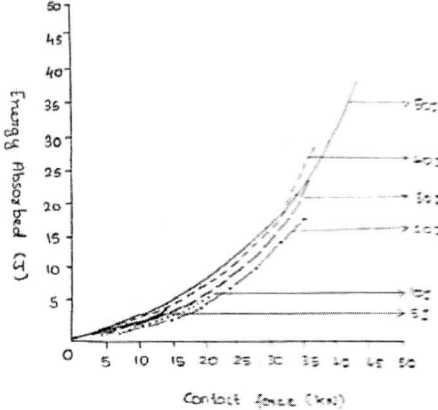

Figure 8 – Graph showing energy absorbed against contact force at various impact energies [20]

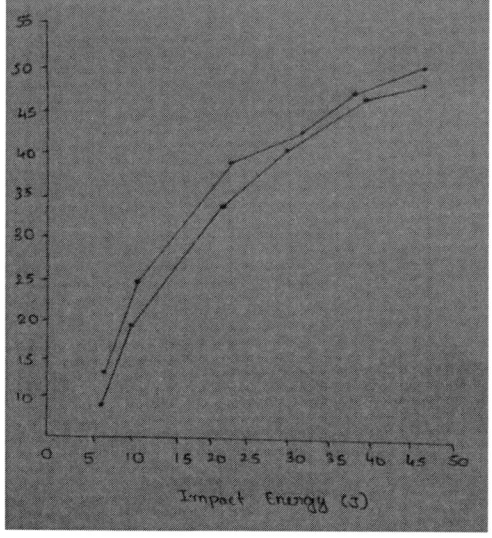

Figure 9 – Variations of the peak contact force and peak transmitted force with impact energy [20]

This is the result of a hole created in the top outer fabric layer when many mono- and multifilaments are broken (Figure 8). Based on the European Standard for motorcyclists limb impact protectors the mean value of the transmitted forces should not be above 35kN and no single value should be above 50kN at an impact energy of 50 J. The results of the findings indicate that a single fabric layer can satisfy the needs of force transmission when the impact energy is below 30 J. Owing to the severity of motorcycle impacts, a single layer of this specific fabric is hence inadequate to offer required protection. Under such condition, the fabric structure should be optimized and lamination used to help the protective performance of spacer fabrics to give the necessary protection. The influence of structural factors and lamination on protective properties shall be explained in subsequent sections of this chapter.

5.5   Analysis of frequency domain

A frequency domain study has been done adopting the Hilbert–Huang transform (HHT) so as to enable a deeper understanding of the impact behavior of the fabric under hemispherical form and to elucidate the failure mechanisms within the time domain. The techniques of frequency domain analysis comprise of Fast Fourier transform (FFT), wavelet transform, HHT, and so on. The generally used technique, namely, FFT, is only able to process linear and stationary signals [13]. Wavelet transform is suitable for non-stationary signals, but not for non-linear signals [14]. HHT is a combination of Hilbert spectral analysis and empirical mode decomposition (EMD) to separate a signal into intrinsic mode functions (IMF) and to obtain instantaneous frequency data [15–17]. It is flexible and proves very efficient for analyzing non-linear and non-stationary data in time–frequency–amplitude representation. It has also been established to have the ability to reveal hidden physical meanings in the data. It has thus been used to carry out the study of frequency domain of the transmitted force signals and establish the frequency features that correspond to the deformation and damage modes of the fabric. The study has been carried out with ensemble EMD (EEMD) that defines the IMF components as the mean of an ensemble of trials, each comprising of the signal and white noise of finite amplitude in order to avoid mode mixing, and hence enable to obtain better results [18]. The Hilbert spectra pertaining to the transmitted force signals at various impact energy levels are obtained in the time–frequency–amplitude representation (Figure 15). It gives a measure of amplitude contribution from every frequency and time. Integration of the Hilbert spectrum in the time domain gives the marginal spectrum, which gives a measure of the total amplitude contribution from each frequency. Considering the transmitted force–time signals (Figure 4(b)), the transmitted force amplitudes have been found to be situated at various frequency bands (figure 15). It indicates various deformation and damage modes at various impact energy levels. Also, the frequency distribution also varies with time. This shows that during the process of impact the deformation mode changes. The damage and deformation modes can be identified by the marginal spectra [19]. A specific deformation or damage mode is related to a range of the frequency distributions. The frequency distributions comprise of a many ranges. In the case of plastic deformation of monofilaments, breakages of monofilaments, breakges and thermoplastic transformation of multifilaments frequency ranges have been specified. When the frequency ranges have been determined, the related deformation and damage modes can also be determined. At specific low level of impact energy the frequencies are situated in defined ranges. But, in addition to these frequency ranges the frequencies distributed in the range 50–100 Hz can also be noticed at a particular value of the impact energy. The highest amplitudes are situated within the range 0–10 Hz for impact energies of either 5 J or 10 J.

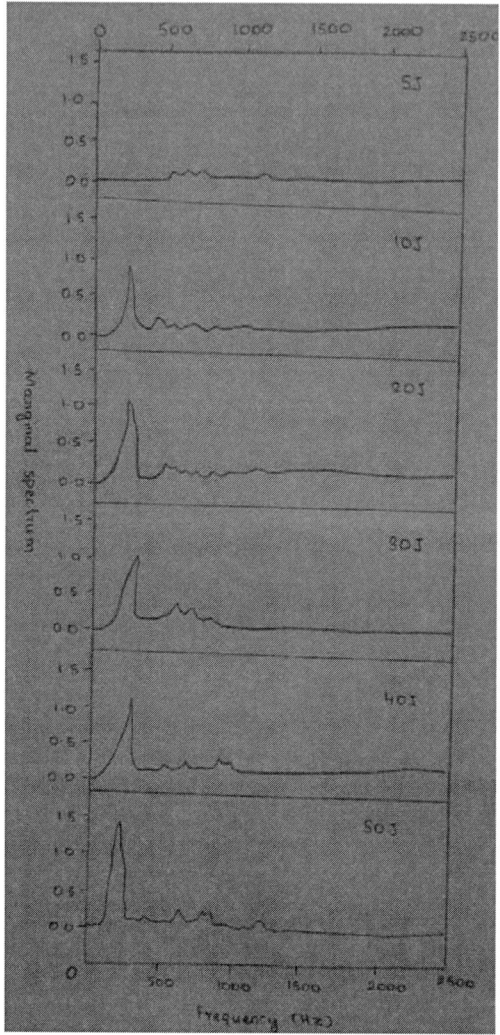

Figure 10 – The marginal spectra for the transmitted force signals derived from various impact energy levels [20]

Based on analysis of frequency pattern, it is possible to evolve a relationship between the frequency ranges and the deformation and damage modes. The plastic deformation can possibly take place between 0–10 Hz and 300– 425 Hz, monofilament breakages take place between 50 and 100 Hz, breakages of multifilaments take place between 80– 180 Hz and 400–560 Hz, and the thermoplastic transformation happens between 180–210 Hz and 560– 700 Hz. Such findings reveal that at greater impact energy, the frequency distribution range in the marginal frequency spectra is higher. Higher frequency ranges as well as higher transmitted forces also relate to the severe damage of the fabric. Thus knowing the frequency ranges the deformation and damage modes in the spacer fabric can be identified from the frequency domain analysis.

5.6   Influence of the inclination of spacer monofilament

In order to study the influence of the spacer inclination on the impact force attenuation properties, the following types of warp knit spacer fabrics having similar outer layer structure (chain plus inlay) and spacer monofilament yarn but with different spacer monofilament inclinations (underlapping one needle, two needles, and three needles between the front- and back-needle bars) have been used

   a) Fabric having inlay structure for top and bottom layers and single lapping of spacer monofilament inclination.

   b) Fabric having inlay structure for top and bottom layers and double lapping of spacer monofilament inclination.

   c) Fabric having inlay structure for top and bottom layers and triple lapping of spacer monofilament inclination.

The thickness and stitch density of the fabric are maintained almost similar in the outer layers. The inclination and length of the spacer monofilament determine the number of the needles underlapped. The spacer monofilaments become longer and more inclined as the number of the needles underlapped increases.

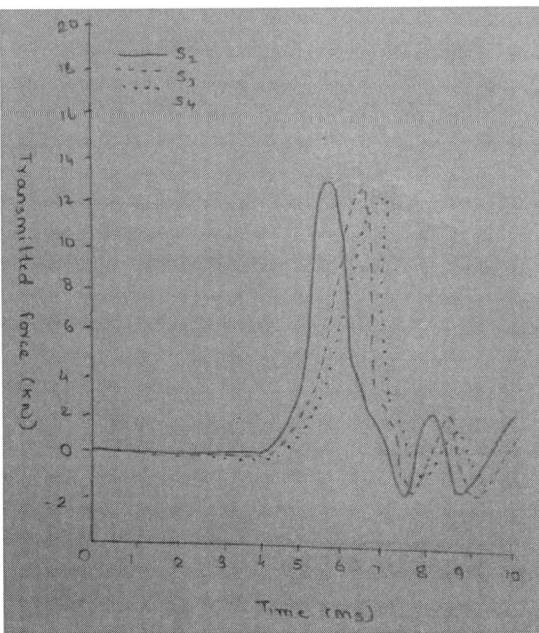

Figure 11 – Influence of inclination of spacer monofilament on transmitted force-time curves [21]

In order to explain the influence of the spacer monofilament inclination having same impact energy the transmitted force-time curves of these fabrics in a single layer under impact at a specific value of impact energy is depicted in Figure 1. It can be seen that while the duration from the beginning point where the striker contacts the fabric upper surface to the peak point where the transmitted force reaches the maximal value increases as the spacer monofilament inclination increases. There is reduction of the peak transmitted force with the increase in the spacer yarn inclination. It implies that the spacer fabric having a greater spacer monofilament inclination and a longer spacer monofilament length can resist the impact due to a lower peak transmitted force in a better way. The finding considerably differs from the flat wise impact test result obtained from earlier work, and has shown that the fabric having moderate inclination spacer monofilaments (two needles underlapped) has the lowest peak transmitted force. Shear can easily take place between the two layers under the flatwise impact as the fabric with too vertical spacer yarns (one needle underlapped) becomes less stable. Whereas, the spacer monofilaments will have a longer length in a fabric, and will be less oriented to the direction of impact compression when they are too inclined (three needles underlapped). A longer elastic rod that is less oriented to the compression direction has a lower critical load based on the theory of elastic stability. Hence, the fabric having moderate inclination of spacer monofilaments (two needles underlapped) would exhibit the best compression resistance to the flatwise impact, and would enable it to absorb more impact energy at the plateau stage. The peak transmitted force is thus the lowest. But, the situation gets complex under impact in the hemispherical shape. In the case of a planar spacer fabric its top outer layer will be extended and bottom layer contracted when it is placed on the hemispherical surface of the anvil. Such special boundary state enables the spacer monofilaments to more easily shear than to buckle under impact. Under such condition, the stability of the spacer monofilaments is important for the spacer fabric to resist the impact, since the shear movements would reduce the compression resistance of the spacer yarns to the impact load, and results in a high peak transmitted force. The peak transmitted force will be lower as the fabric having greater inclination spacer monofilaments has a higher shear resistance. Hence, considering the impact in hemispherical form among the three types of fabrics considered above, the force attenuation performance of the fabrics increase with the increase in the length and inclination of spacer yarns. The peak transmitted forces of spacer fabrics considered under impact with various kinetic energies and laminated layers have been determined.

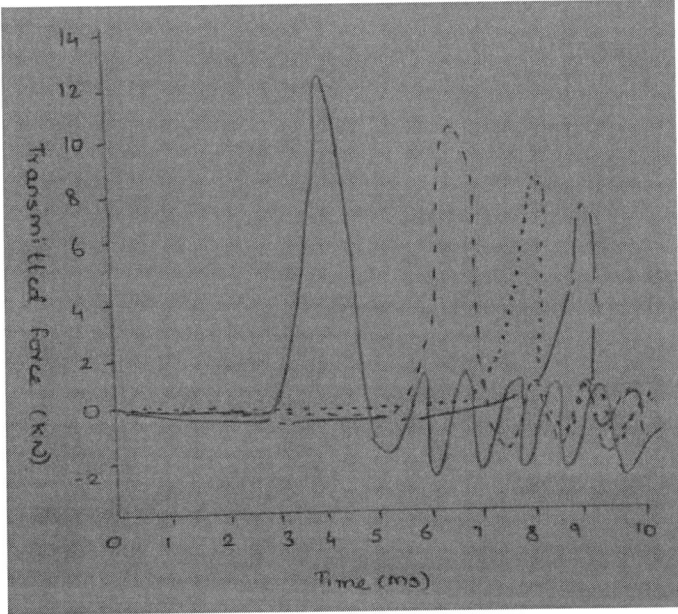

Figure 12 – Influence of spacer monofilament diameter on transmitted force-time curves [21]

The same trend, wherein the peak transmitted force reduces as the spacer monofilament inclination increases, is noticed for other levels of impact energies and laminated layers. But, the peak transmitted force values reduce drastically with the increase in the number of fabric layers. The influence is not so clear in a single layer, despite the reduction in the peak transmitted force with the increase of the spacer monofilament inclination. But, the influence of the inclination of the spacer monofilament becomes very significant by increasing the fabric layers. It is so since a single layer of the spacer fabrics is not strong enough to resist the impact with a high kinetic energy and is easier to be compressed into a high densification stage. The 3 types of fabrics have been compressed into their high densification stage in a single layer. The three types of fabrics have been compressed into their high densification stages. Under such situations, the striker compressed the monofilament material rather than making the spacer monofilaments buckle. As discussed earlier, the transmitted force is based on the contact stiffness of the compressed yarn material. As the fabrics have been produced from polyester monofilaments, the difference in peak transmitted force among fabrics is insignificant. But, there is increase of the absorption of energy of the laminated fabrics with the increase of the fabric layers. Hence, the fabrics will not be compressed into a high densification stage. Under such situation, the post-buckling of spacer monofilaments becomes crucial in resisting the impact loading and the spacer monofilament inclination effect works. Hence, in the case of fabrics having more layers, a high difference in the value of the peak transmitted force can be seen. Another consideration is that by increasing the fabric layers, better impact force attenuation of spacer fabrics can be obtained when the impact energy is at a lower level. For example, addition of a single fabric layer can almost decrease the peak transmitted force by 50% under impacts with the kinetic energies of a certain value and double its value.

5.7   Influence of the fineness of the spacer monofilament

In order to investigate the fineness of the spacer monofilament on the impact force attenuation properties of warp knit spacer fabrics, two fabrics having the same spacer monofilament inclination (underlapping three needles) and the same outer layer structure (chain plus inlay) but with two different spacer monofilament diameters (0.2 and 0.16 mm) have been selected. The stitch density of the outer layer as well as thickness of the two fabrics has been maintained almost equal. The transmitted force–time curves of these two fabrics in a single layer under impact at a chosen value of kinetic energy have been used as an example to study the influence of the spacer monofilament fineness with the same impact energy. The spacer fabric having the coarser spacer monofilament has a lower peak transmitted force and a longer time to the peak point and, hence, possesses a better impact force attenuation property. This result is consistent with that obtained from the flatwise impact test, although the boundary condition is changed to the hemispherical form. It is considered normal since the fabric having coarser spacer monofilaments exhibit higher compression resistance which can decelerate the striker more quickly and make the striker experience a longer time to reach the peak transmitted force point than the fabric with finer spacer monofilaments. For decreasing the peak transmitted force the impact process should be as long as possible for absorbing more energy as the duration of deceleration is necessary for the impact protection. This finding is also valid for other impact energy levels and fabric laminated layers. Just as in the case of the flatwise impact, the force attenuation properties of spacer fabrics in the hemispherical shape can be improved considerably by increasing the spacer monofilament diameter. Despite increasing the spacer monofilament diameter can considerably improve the force attenuation capacity of the spacer fabric in the hemispherical shape, increasing the spacer monofilament diameter can increase the fabric stiffness and, thus, decrease the comfort property of the fabric. Therefore, it is necessary to consider the balance between the comfort and protective performance through choice of a proper spacer monofilament fineness for a particular protective application.

5.8   Influence of the fabric thickness

In order to study the influence of fabric thickness on the impact force attenuation properties of warp knit spacer fabrics in the hemispherical form, four types of warp knit spacer fabrics having similar monofilament diameter, spacer monofilament inclination (underlapping two needles) and outer layer structure(chain plus inlay) of various thicknesses

Figure 3 depicts the transmitted force–time curves of such fabrics in a single layer under impact at a specific value of kinetic energy which has been chosen for illustration for studying the influence of the fabric thickness having similar impact energy. There is reduction in peak transmitted force and with the increase in the thickness of the fabric the duration from the starting point to the peak transmitted force point increases. The process can be described as below

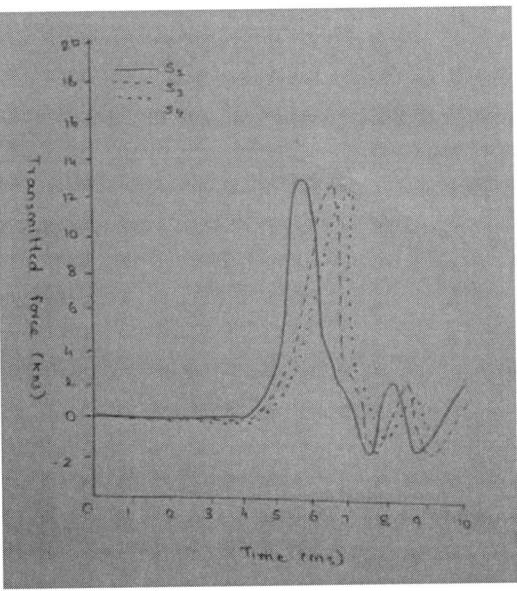

Figure 13 – Influence of thickness of fabric on transmitted force time curves [21]

Firstly, more time is necessary to compress a thicker fabric to its densification stage at a larger displacement. In the case of a thicker fabric it is possible to achieve a lower peak transmitted force, as the increase of the compression time and displacement permits a thicker fabric to absorb more impact energy. Secondly, with the increase in striker displacement the contact area of the fabric with the striker rapidly increases. Many spacer monofilaments are involved in resisting impact loading since a thicker fabric can be compressed into a larger displacement. Hence, in the case of impact in hemispherical form the thicker fabric has a lower peak transmitted force. The peak transmitted forces for various impact energies and laminated layers have been determined. The similar trend, in which the peak transmitted force decreases as the fabric thickness increases is also obtained for other impact energy levels and laminated layers when all the fabric samples have the same destruction modes under impact. But, there are two exceptions for a single layer of the fabrics under a certain value of impact and double layers of the fabrics under impact with certain value. In comparison with other fabrics, the thickest fabrics In these two special cases, the thickest fabrics do not exhibit the lowest peak transmitted forces because of the various destruction modes, considering these two special cases. The damage of multifilaments in the top outer layers of fabrics in a single layer can be seen under impact at a certain value of kinetic energy, in single layer can be observed. However, no obvious damage can be seen in the thickest fabric. At a particular value of impact similar phenomenon is seen in double layers of the fabrics. The peak transmitted forces of the thickest fabric can be higher than that of a thinner fabric as it cannot absorb additional impact energy because of the damage of the fabric structure in these two cases. It can be seen from the aforesaid discussions that there needs to be a proper choice of the fabric thickness based on the destruction modes when a spacer fabric will be subjected to impact with higher levels of kinetic energies during use.

5.9   Influence of the outer layer structure

In order to study the influence of the outer layer structure on the impact force attenuation properties of warp knit spacer fabrics a number of fabrics having similarly underlapped needles for the spacer yarns (two needles) and almost similar thickness but with different outer layer structures have been selected. In order to study the influence of the outer layer structure on the impact force attenuation properties, the transmited force-time curves of such fabrics in a single layer under impact at a kinetic energy of specified value has been used for illustration(figure 4). The outer layer structures have a definite influence. On the one hand, fabric having large size hexagonal meshes for both outer layers has the highest peak transmitted force and the shortest duration from the starting point to the peak transmitted force point. On the other hand fabric having a chain plus inlay structure for both outer layers has the lowest peak transmitted force and the longest impact duration. In the case of fabrics with other outer layer structures the peak transmitted forces lie between the values for these two fabrics. With other levels of impact energies and laminated layers, similar results have also been obtained. Such differences arise chiefly from various geometric features of outer layer structures that result in various geometric arrangements of multifilaments and various inclinations and binding conditions of spacer monofilaments. As shown in earlier investigations under flatwise impact, the fabric having large-size meshes show the poorest impact protective performance because of highly buckled and inclined spacer monofilaments, and the fabric knitted with small-size meshes exhibits the best impact protection ability because of tight binding conditions. Because of the combined influences of loose binding structures and lowly buckled and inclined spacer monofilaments the fabrics having a close structure exhibit moderate impact protection performance. In the case of impact in the hemispherical form, besides the above stated factors, the deformation of the outer layer structures to fit the shape of the anvil also considerably affects the impact force attenuation properties of the fabrics. It is clear that an open or mesh structure is less stable and stiffer in comparison with a close structure. In the case of the close structures, the chain plus inlay structure has been found to be more stable out-of-plane in comparison with the locknit structure, since the chain loops and the inlay yarns are crossed in a perpendicular manner which are more difficult to extend and shear than the locknit stitches (Figure 5). In comparison with a fabric having a less stable outer layer structure, a fabric having a more stable outer layer structure can be less sheared and extended, as it is placed onto the hemispherical surface of the anvil. Hence, owing to a lower deformation of the stable outer layer structure the number of spacer monofilaments which can resist the impact cannot be significantly reduced due to a lower deformation of the stable outer layer structure. Whereas, the stable outer layers render the fabric stiffer, and thus enables to disperse the stress wave to a greater area and absorb more energy. Thus, the best force attenuation performance is achieved with the fabric knitted with the most stable outer layer structure (chain plus inlay). The fabric having large-size hexagonal meshes exhibit the poorest impact force attenuation due to the combined effects of low stable outer layer structure and highly buckled and inclined spacer monofilaments.

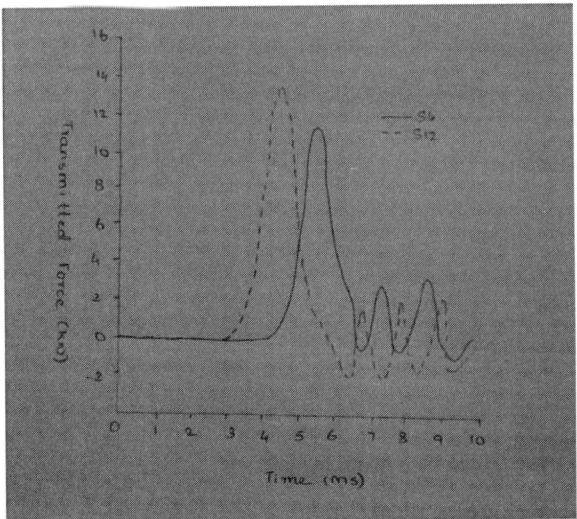

Figure 14 – Influence of outer layer structure on transmitted force time curves [21]

From the above aforesaid discussions it is clear that influences of outer layer structure on the force attenuation properties under impact in the planar form differs from that of the hemispherical form. The comfort and formability of fabrics stand in contradiction to the impact force attenuation properties. In comparison with a fabric having flexible structure the fabric having a stable and stiff outer layer structure exhibits better force attenuation ability. But, the stiff outer layer structure renders the fabric uncomfortable and hard to fit a curved shape. Hence, further studies are necessary to find the balance between the comfort and protective performance in order to choose appropriate outer layer structure for required end uses.

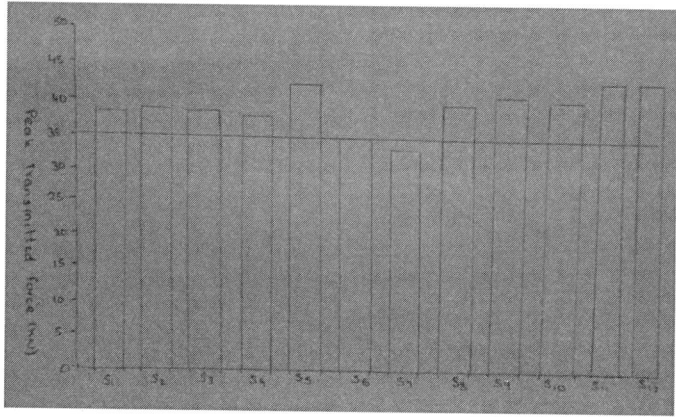

Figure 15 – Peak transmitted force of spacer fabrics laminated with three layers [21]

Studies have been carried out on the effects of the structural factors, including spacer monofilament inclination and fineness, fabric thickness, outer layer structure and fabric lamination on the impact force attenuation properties of the warp-knitted spacer fabrics. The findings have revealed that such structural factors considerably influence the resultant peak transmitted force. Thus, in order to achieve wearing comfort a warp-knitted spacer fabric can be designable by maximizing its force attenuation capacity and also minimizing its density and thickness. The findings have also revealed that the structural factors do not determine the peak transmitted force independently. The peak transmitted force is also affected by the predefined impact kinetic energy. Hence, no best spacer fabric exists for all impact kinetic energy levels. It is initially necessary to identify the particular application with an impact kinetic energy level in order to optimize the structure of spacer fabrics. Subsequently, it is possible to design appropriate spacer fabrics by varying their structural factors quantitatively to achieve a specific peak transmitted force under a particular impact kinetic energy. As the human body can undergo various levels of energy impacts in various situations or sports, different kinds of impact protectors have been commercially available for human body protection. Recently, the use of the European Standard for evaluation of motorcycle protective clothing has become widely popular. The protective performance of warp knit spacer fabrics have been compared with required standards so as to assess the validity of replacing commonly used polymeric foams with the developed spacer fabrics. Based on the standard, the peak transmitted force of limb protectors for motorcyclists should not be above 35 kN, and no single value should be above 50kN under impact at a kinetic energy of 50 J. The peak transmitted forces of all the developed spacer fabrics laminated with three layers is depicted in figure 6. Only two types of tested fabrics among others are found to comply with the standard. These two spacer fabrics are knitted with the same outer layer structure (chain plus inlay), but with different thicknesses and different stitch densities. Even though one fabric has a smaller thickness than the other, three layers of the first one still have a thickness of 25.35 mm, which is too thick for use in protective clothing. As a fabric having coarser and more inclined spacer monofilaments exhibits a better force attenuation capacity, the thickness can be reduced by increasing the diameter or inclination of the spacer monofilaments. But, increasing the spacer monofilament diameter will render the fabric stiffer, leading to a decrease in comfort. In order to obtain a good balance between the force attenuation capacity and comfort it is necessary to further design and optimize the spacer fabrics by manipulating the structural factors

5.10   References

[1]   Viano DC, Bir CA, Cheney AK, et al. Prevention of commotion cordis in baseball: an evaluation of chest protectors. J Trauma 2000; 49: 1023–1028.

[2]   Rønning R, Rønning I, Gerner T, et al. The efficacy of wrist protectors in preventing snowboarding injuries. Am J Sport Med 2001; 29: 581–585.

[3]   Lemair M and Pearsall DJ. Evaluation of impact attenuation of facial protectors in ice hockey helmets. Sports Engineering 2007; 10: 65–74.

[4]   Bernhardt DT. Protective sports equipment. In: Hebestreit H and Bar-Or O (eds) The young athlete. New York: Wiley, 2008, pp.164–168.

[5]   Parker MJ, Gillespie WJ and Gillespie LD. Effectiveness of hip protectors for preventing hip fractures in elderly people: systematic review. BMJ 2006; 332: 571–574.

[6]   van Schoor NM, van der Veen AJ, Schaap LA, Smit TH and Lips P. Biomechanical comparison of hard and soft hip protectors, and the influence of soft tissue. Bone 2006; 39: 401–407.

[7]   Laing AC, Feldman F, Jalili M, Tsai CM and Rabinovitch SN. The effects of pad geometry and material properties on the biomechanical effectiveness of 26 commercially available hip protectors. J Biomech 2011; 44: 2627–2635.

[8]   Bellfy PI. Attachment of protective pads for protection of joint surfaces. US Patent 7487557, 2009.

[9]   Dlugosch S, Hu H and Chan CK. Impact protective clothing in sport: areas of application and level of utilization. RJTA 2012; 16: 18–28.

[10]   Liu YP, Hu H, Zhao L, et al. Compression behavior of warp-knitted spacer fabrics for cushioning applications. Text Res J 2012; 82: 11–20.

[11]   Liu YP, Hu H, Long HR, et al. Impact compressive behavior of warp-knitted spacer fabrics for protective applications. Text Res J 2012; 82: 773–788.

[12]   Guo XF, Long HR and Zhao L. Investigation on the impact and compression-after-impact properties of warp-knitted spacer fabrics. Text Res J 2013; 83: 904–916.

[13]   Laing AC, Feldman F, Jalili M, et al. The effects of pad geometry and material properties on the biomechanical effectiveness of 26 commercially available hip protectors. J Biomech 2011; 44: 2627–2635.

[14]   Schmitt KU, Nusser M, Derler S, et al. Analysing the protective potential of padded soccer goalkeeper shorts. Br J Sports Med 2010; 44: 426–429.

[15]   Melissa AS, Dennis JC and Rebekah OH. Injury prevention in sports. Am J Lifestyle Med 2010; 4: 42–64.

[16]   Liu YP, Hu H, Zhao L, et al. Compression behavior of warp-knitted spacer fabrics for cushioning applications. Textile Res J 2012; 82: 11–20.

[17]   Liu YP, Hu H, Long HR, et al. Impact compressive behavior of warp-knitted spacer fabrics for protective applications. Textile Res J 2012; 82: 773–788.

[18]   Liu YP, Au WM and Hu H. Protective properties of warp knitted spacer fabrics under impact in hemispherical form. Part I: impact behavior analysis of a typical spacer fabric. Textile Res J. Epub ahead of print. DOI: 10.1177/ 0040517513495941.

[19]   BS EN 1621-1:1998: Motorcyclists' protective clothing against mechanical impact. Requirements and test methods for impact protectors.

[20]   Yanping L, Wai M A and Hong H, Protective properties of warp-knitted spacer fabrics under impact in hemispherical form. Part I: Impact behavior analysis of a typical spacer fabric, Textile Research Journal, 2014, 84(4) 422.

[21]   Yanping L, Wai M A and Hong H, Protective properties of warp-knitted spacer fabrics under impact in hemispherical form. Part II: effects of structural parameters and lamination, Textile Research Journal, 2014, Vol 84(3) 312

Chapter 6

Spherical compression properties of warp knit spacer fabrics

Summary

Warp knit spacer fabrics have been theoretically evaluated for spherical compression. The evaluation has been based on the fundamental assumptions derived from a preliminary test. A relation has been established between the total compression force applied by the spherical ball, and compression strain at the point of highest compression by analyzing the shape of the fabrics deformed under the ball compression. Considering the introduction of non dimensional parameters, the influences of both ball radius and fabric thickness have been explained. The findings reveal that the spherical compression effects increase inversely with increasing the ball radius, and increase directly with the fabric thickness. The theoretical analysis has been followed up by comparison experimental results with theoretical calculations. Using plane plate as well as spherical ball compression conditions, various knit spacer fabrics have been tested. There are three areas in the plane compression stress-strain curves of the fabrics. A linear equation nearly represents the curve segment in each area/region. In the case of each fabric, the compression curve under spherical ball compression is calculated and compared with the experimental curve depending on the constants E and s0 obtained. Three compression parameters, including the maximal compression force and compression work at the compression strain of 0.70, as well as the linear degree, are also used for comparison. The theoretical calculations agree well with the experiments. The spherical compression behaviors of knitted spacer fabrics compressed with various radii of the balls from the plane compression testing results can be predicted with the help of the validated theoretical model.

6.1   Introduction

A number of special textile products have been produced from warp knit spacer fabrics by altering their structure design and finishing methods to suit broad areas of applications that include sound absorption, moisture transport, functional bra support, comfort property enhancement, car seats and composite reinforcement [1-8]. Their compression properties mainly decide the areas of applications, since they have a sandwich type of structure. The compression properties of spacer fabrics intended for use as apparels requires soft handle and also enable to fit the body shape. Warp as well as weft knitted spacer fabrics can be designed as a type of soft spacer fabric to satisfy the requirement of special uses like bra cups and flexible cushioning or padding materials [9]. They can be constructed as two separate knitted fabrics as the top and bottom layers and flexible filaments as the centre layers. Various knitted spacer fabrics intended for intimate apparel have been studied experimentally with regard to physical and compression properties under low stress [10,11]. Compared to polyurethane foams, knit spacer fabrics show better pressure distribution, air permeability and heat resistance as pointed out by other investigations [7,9]. The knit spacer fabrics can specifically be used as pressure release products like functional mattresses and wheelchairs in order to decrease peak pressure for preventing the concentration of pressure on the body [12-19]. In order to effectively evaluate the applications of the knitted spacer fabrics in different shapes of body, it is important to understand their compression properties under spherical compression conditions. Though the compression properties of knit spacer fabrics have been subjected to many experimental and theoretical studies they have merely concentrated on the compression behavior under plane compression conditions. The spherical behavior of warp knit spacer fabrics has been reported only in a particular work [20]. This chapter firstly highlights the spherical compression tests of different warp-knitted fabrics and the compression stresses at the maximal compression point of the spherical ball for different compression strains have been calculated based on the Hertz theory developed for elastic homogeneous materials under small strain conditions [21]. Despite including a finite element simulation here, analytical models for predicting the spherical compression behavior of knitted spacer fabrics are missing. The spherical compression behavior of knitted spacer fabrics have been discussed herein. The chapter discusses the findings in two parts. The first part concentrates on the theoretical analysis of a spacer fabric under spherical ball compression. The spherical ball compression properties of the fabric can be predicted by a theoretical model based on the compression constants derived from its plane compression stress-strain curve. The influences of the fabric thickness and ball radius can be explained by non-dimensional parameters, which are derived from the theoretical relation between the compression force and compression strain at the point of highest compression. The investigation would enable better knowledge of the behaviors of knitted spacer fabrics under spherical compression.

During the first part of investigation the spherical compression properties of knitted spacer fabrics have been predicted through elaborate theoretical analysis to establish a theoretical model [22]. In this model, the compression stress–strain curve of a knitted spacer fabric under the plane compression condition has to be used to derive the compression constants that are required for the theoretical calculation of its compression force–strain curve under the spherical compression condition. Based on earlier investigations the compression stress–strain curve of a knitted spacer fabric can be divided into various zones, and a linear equation can be used to represent the compression stress– strain curve at each compression zone [23-27]. The first part of the investigation deals with the theoretical analysis and model followed by the comparison of the model with experimental results. Linear fitting techniques have been used in the plane compression stress–strain curves corresponding to various spacer fabrics which are first provided by the experiments to calculate the modulus and intercept. Then, the modulus and intercept obtained are used to calculate the total spherical compression force and compression strain curve for each fabric sample. The model is validated lastly by comparing the calculated curves with experimental ones. Depending on the compression factors chosen the differences between the calculations and experiments have also been discussed.

6.2   Derivation of the model

Details of compression methods

Firstly the compression methods have to be briefly discussed so as to make the basic assumptions for a theoretical model. Two compression methods can be switched by just changing the corresponding compression assemblies. Figure 1a and b depict the plane compression test the spacer fabric which is compressed between two plane plates (one fixed and the other movable), and the spherical compression test that is conducted by moving a spherical ball to compress the knitted spacer fabric which is placed on the fixed platform, respectively. The lower face of the fabric has to be secured to the platform with adhesive tape so as to prevent its possible movement during the compression test. The shape of the knit fabric at the contact area with the spherical ball is considered crucial and has to be determined first by experiment so as to carry out a theoretical analysis of its spherical compression behavior. Since an isotropic material has similar property in all directions, its contacting shape should be circular. But, owing to its anisotropic behavior, the condition is different in the case of spacer fabric. An initial test has been carried out so as to obtain the contacting shape of the knitted spacer fabric with the spherical ball surface.

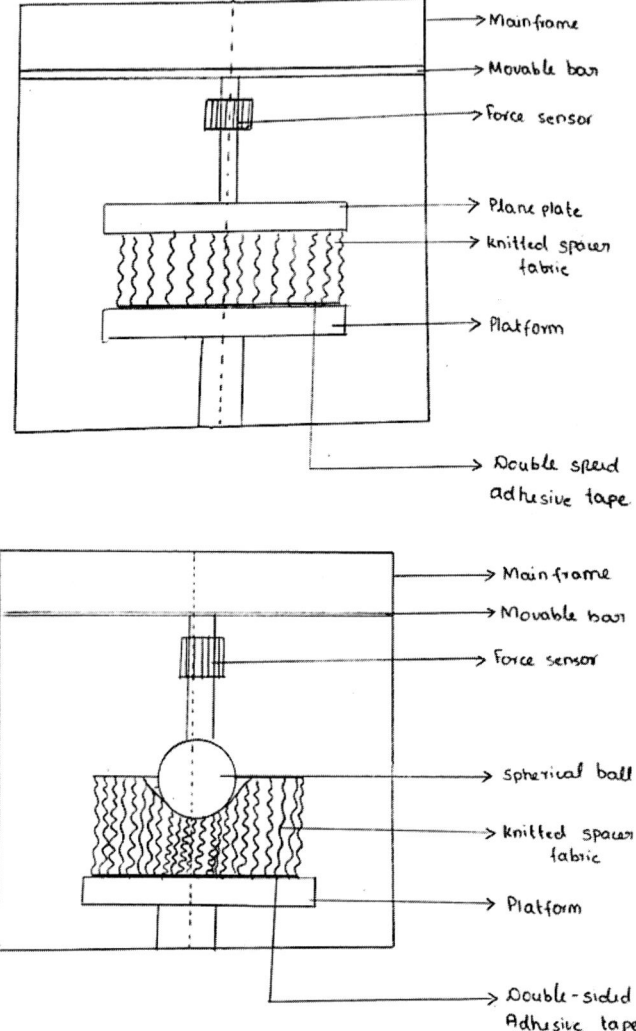

Figure 1 - Compression testing methods shown schematically [28].
a)   Plane plate compression
b)   Spherical ball compression

The contacting shape of the fabric at a given displacement of the spherical ball was obtained by coating powder on the knitted spacer fabric face (figure 2). It can be found that the contacting contour is almost circular.

The following assumptions are made for the theoretical analysis of the spacer fabric under spherical compression considering the change in fabric shape noticed during the initial test:

a) The spacer filament yarns bear the compression force applied by the spherical ball. When undergoing compression test, there is no change in the vertical position of the spacer yarns along the direction of compression. Such assumption depends on the fact that no obvious movement of the specimen has been noticed during the spherical compression test because of the fixation of the bottom side of the specimen to the platform, and there is no change in the vertical direction of two ends of spacer filaments in the inner layer. Hence, the component forces of spacer filaments in the horizontal direction in the inner layer can be ignored.

b) The spacer filaments in the inner layer are assumed as spring-like elements under compression. By application of the compression forces at either side, they are deformed between the upper and bottom layers. In a state of compression, the deformations of a spacer filament usually comprise of axial compression, bending and torsion. There is a resultant influence in all such deformations to support compression deformation of the fabric. The investigation the plane compression behavior of the idealized spacer yarns has been used so as to avoid the complex analysis of a single spacer filament under compression. This can simplify the analysis. Based on the fabric plane compression properties the idealized spring-like filaments can be linear or non-linear.

c) The top layer of the fabric can be split in two areas under the spherical compression, namely, the area in contact with and without the ball surface (figure 9). The contour of the contacting area is assumed to be a circular form, and the expanding curve in the non-contacting area is assumed to have a shape of an exponential function whose asymptotic line is through the upper surface of the knitted fabric. On the other hand, circular contacting curve and the expanding curve have the same slope at the point of separation between the spherical ball and the knit fabric.

d) The total compression force exerted by the ball on the fabric is the product of the sum of the resultant forces withstood by both the spacer yarns in the contacting area and non-contacting area along the plane of compression. The influence non-contacting area along the plane of compression is transferred to the spherical ball through the deformation tension of the top layer of the fabric in the non-contacting area.

6.3   Study of the geometry

Considering the Figure 2, the following parameters are used

T(H1) – the fabric thickness

D(H2) – the displacement of the ball

R(R0) – the radius of the ball;

R1(r0) – the radius of the ball at the separating point

θ (0) – the angle formed between the central line and at the separating point.

Referring to Figure 2 and using third assumption, it is possible to find out the shapes of the upper layer on the cross-section in both the contacting and expanding areas. The area of contact is spherical shape since the fabric is contacted with the surface of the ball. Considering that the vertical direction line passing through the spherical center point and the contacting area between the knitted spacer fabric and spherical ball surface is a spherical surface.

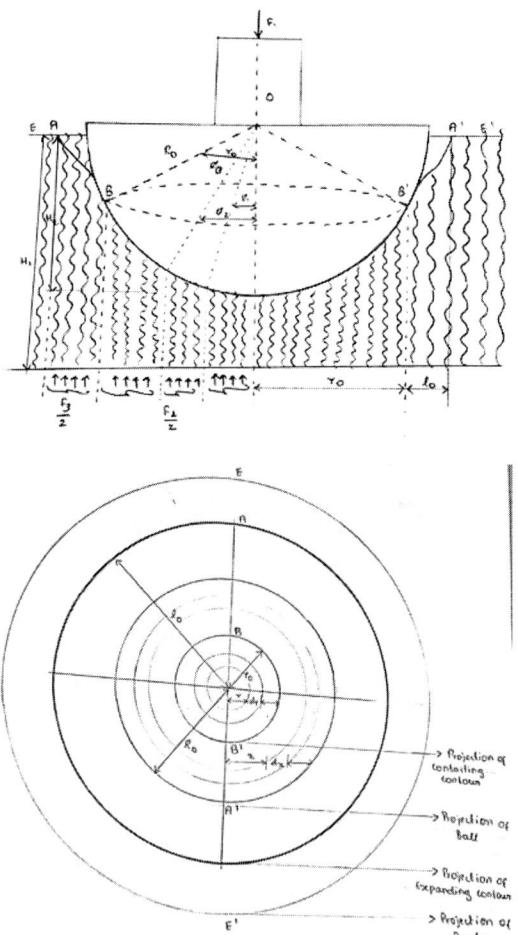

Figure 1 – Compression of spherical ball [28].
a)   Cross section view
b)   Projection view

The non-contacting area is more complex than the contacting area and is considered as the expanding area. The asymptotic line is located on the upper surface of the knitted spacer fabric. For subsequently enabling the calculation of points that are very close to the upper surface of the fabric have been selected as the limit points, and the horizontal distance a point to another is defined as the expanding span width. In order to derive the function of the expanding curve starting from a particular point a cartesian coordinate system is established by setting horizontal line as the x-axis and vertical line as the y-axis. Depending on the third assumption, an exponential function is proposed to be a curve function of whose asymptotic line is set as a line.

The ball radius is not zero as the balls compression displacement is lesser than the spherical ball diameter.

It has been noticed that during the spherical compression test, the expanding zone span width in the expanding area under compression deformation did not exceed the border of the knitted spacer fabric sample.

In the study, the coefficient has been chosen as 0.99995, which corresponds to a compression strain level less than 0.005% from a certain point to the border of the knitted spacer fabric. Under such level of deformation, the resultant force withstood by the spacer filaments in the expanding area is less than 0.05% of the resultant force withstood by the spacer yarns in the expanding area. Under this situation, only the compression force in the expanding area from certain two different points has been taken into account. Since the effect is very small the effect from the reference points to the border of the fabric can be neglected.

6.4   Study of compressive force

Based on the fourth assumption, the total compression force applied by the spherical ball on the fabric equals the sum of the resultant forces withstood by the spacer filament yarns in the contacting area as well as non-contacting area along the compression direction. Thus, the total compression force can be expressed mathematically.

Based upon the equilibrium force, the resultant of forces withstood by the spacer yarns in the contacting area and the resultant of forces withstood by the spacer yarns from the expanding area is equal to the reaction of resultant forces applied by the platform on the bottom surface of the fabric. The contacting area and the expanding area correlate with the compression strain of the knitted spacer fabric which is varied along the upper layer of the fabric for a given displacement of the compression ball.

In order to find out the reaction force/unit area on the bottom surface of the fabric, the above study reveals that it is necessary to know the calculations of contacting and expanding areas of the fabric. Depending upon the first and second assumptions can be calculated from the plane compression testing results for the same compression strain. Figure 2 depicts the plane compression of a knitted spacer fabric, where a plane specimen having a specified thicknessand a surface is compressed to a certain displacement. Earlier investigations have revealed that the compression curve of knitted spacer fabric under plane compression can be split into various regions [18,19].

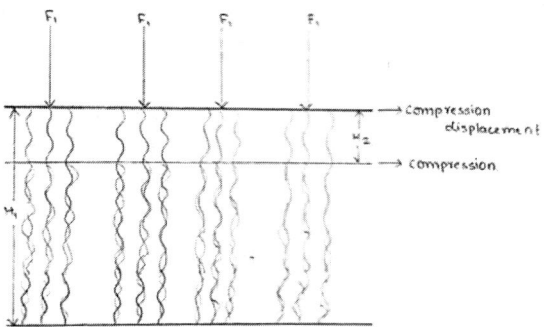

Figure 2 - Spacer fabric under plane compression [28].

Figure 3 depicts typical compression curve of a warp-knitted spacer fabric. The entire curve is divided into three zones and the curve segment in every zone can be approximately represented by a linear relationship between the compression stress and strain. Hence, a relation seems to exist between the compression stress and strain. The resultant of forces withstood by the spacer yarns in the contacting area and the resultant of the forces withstood by the spacer yarns in the expanding area can be calculated by knowing the force of compression strain. Equations for calculating the resultant of forces withstood by the spacer yarns in the contacting area and the resultant of the forces withstood by the spacer yarns in the expanding area have been derived.

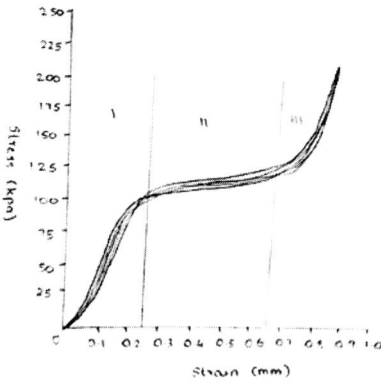

Figure 3 – Curve of compression in a spacer fabric [28].

6.5   Factors considered in the model

A number of non dimensional factors have been considered for providing a better understanding of the spherical compression behavior of the knit spacer fabrics

    a)   Compression strain at the maximum compression point

    b)   Compression force

    c)   Compression force in non dimensional form

    d)   Fabric thickness

    e)   Spherical ball radius

The compression stress strain curve under plane compression is assumed to be linear for each compression stage depending on the derived equation. The compression has been expressed in non dimensional form through a combination of equations. The compression strain in non dimensional form depends on the spherical ball radius and the fabric thickness. Figure 4 depicts the variations of compression force in non dimensional form against fabric thickness different spherical ball radii are depicted. It implies that as the ball radius increases the influence of the ball radius decreases. It is understandable since with increase of ball radius the difference between the spherical and plane compression gets reduced. In fact, the plane compression can be considered as a special case when the spherical ball radius tends to be infinite.

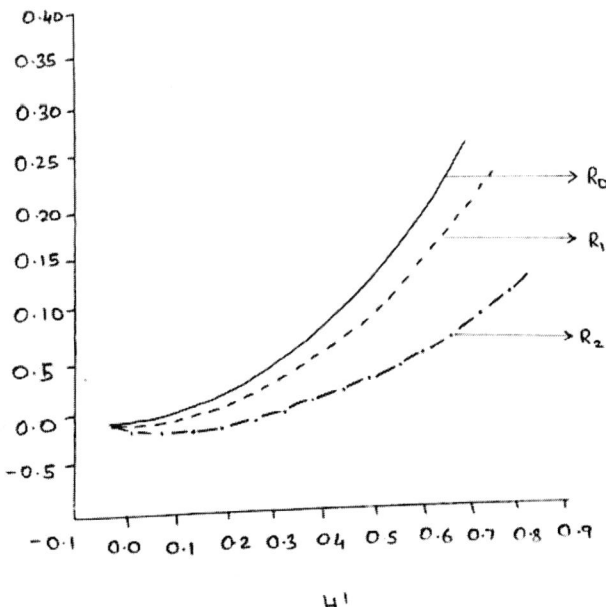

Figure 4 – Variations of F_1 vs H [28].

Figure 5 depicts the variations of the compression force in non dimensional form against the compression strain for different fabric thicknesses when the spherical ball radius is maintained constant. As the fabric thickness increases the compression force in non dimensional form increases, which implies that the compression effect increases as fabric thickness increases. This is normal since increase of the fabric thickness causes the ball contacting the surface with the fabric to increase resulting in a higher influence under spherical compression.

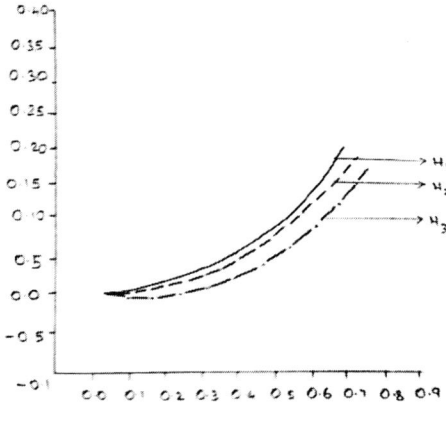

Figure 5 – Variations of F_1 vs H [28].

6.6   Studies on compression

The compression properties have been studied using warp-knitted spacer fabrics having various fabric thickness, surface structure and spacer yarn fineness. The compression behavior of knit spacer fabrics are known to be influenced by such factors. The choice of such fabrics having various structural factors enables a more comprehensive investigation. The details of fabrics tested are shown in Table 1.

Table 1 – Knit spacer fabric details [28]

Fabric	Fabric Thickness (mm)	Outer layer structure	Spacer yarn diameter(mm)
Fabric 1	7.5	Locknit	0.20
Fabric 2	7.5	Chain plus inlay	0.20
Fabric 3	6.8	Hexagonal mesh	0.10
Fabric 4	18	Hexagonal mesh	0.12
Fabric 5	10	Hexagonal mesh	0.22

The fabrics have been produced using polyester filaments. Instron tester has been used for testing the plane as well as plate and spherical ball compression. A pair of circular plates having diameter of 15 cm has been used for the plane plate compression with a compression speed of 12 mm/min. In the case of the ball compression, a fixed circular plane plate of specific diameter and a moveable spherical ball of specified diameter have been used having the same compression speed. The fabrics meant for the plane compression as well as spherical compression have been cut into circular shape having specific diameter. The compression tests have been carried out to the same maximal compression strain, which was 0.70.

6.7   Stress strain curves of plane compression and linear fit

Figure 6 depicts the typical plane compression stress–strain curves of the test spacer fabrics.Under the plane plate compression the compression stress–strain curves are found to be. However, they comprise of three areas (Figure 7). In the first and third areas, as the compression strain increases, the compression stress increases rapidly. In the second area, there is an obvious plateau for the test spacer fabrics1, 2, 3 and 5. Although in the case of test spacer fabric 4 the plateau is not very obviously noticed and its compression curve could be represented by a straight line, the curve is still divided into three linear areas for a better fit. The curve segment in each area can be approximately represented by a linear equation. As can be seen in Figure 3 the compression stress–strain curve is split up into three linear areas (I, II and III) by two critical compression strains. As depicted in figure 2 by linear fitting of the compression stress–strain curves, the moduli, intercepts and critical compression strains for each test spacer fabric are obtained. Also, all the correlation coefficients of linear fitting are greater than 0.95.

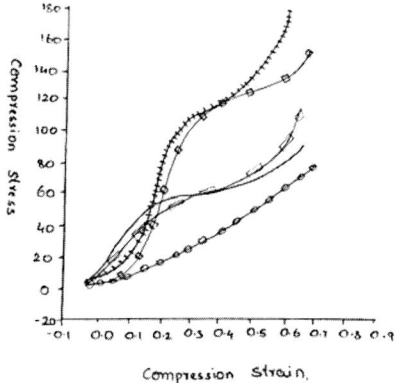

Figure 6 – Plane compression stress strain curves of warp knit spacer fabrics [29].

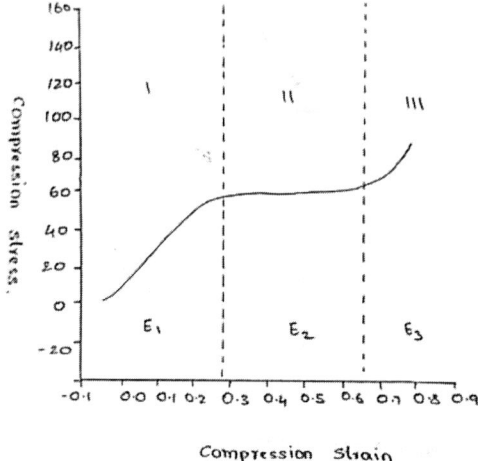

Figure 7 – Plane compression represented by means of straight line [29].

6.8   Calculation of spherical compression curves

Using the data relating to stress strain curves, the spherical compression curve for each knitted spacer fabric can be calculated according to the derived equations. Figure 8 indicates the following parameters

The total compression force applied on the fabric by the spherical ball (F1)

The resultant force withstood by spacer filaments in the contacting area (F2)

The resultant force withstood by spacer filaments in the expanding area (F3)

The compression displacement (H2)

The fabric thickness (H1)

The curvature radius of the spherical ball(R2)

The separating angle between the ball surface and knitted spacer fabric (θ), and

The expanding zone span width of the compression deformation (lo)

During the calculation certain parameters in equations should be replaced by other parameters during the calculation. The choice of required parameters is based on the area in which compression strain is situated. Hence, the critical compression strains can be used to determine the intervals of deformation areas. Based on the previous analysis, the compression strain at any point on the upper surface of the fabric in the expanding area is given by an equation. For the investigation, the compression strain at the maximal compression point has been chosen to be 0.7. At this point, the value of maximum compression strain at the separating point for each test spacer fabric can be calculated. Substituting values in the equation, the maximal compression strain in the expanding area can be obtained. It is found that the values of maximum compression strain in the case of all the test spacer fabrics do not exceed those of strain in first region. It indicates that the compression strains of all the fabrics in the expanding area lie in first region. Thus only moduli in the first region of the compression strain curve and stress are used for replacing moduli and stress in derived equation for calculation of the resultant force withstood by spacer filaments in the expanding area. But, the calculation of the resultant force withstood by spacer filaments in the contacting area for the contacting area is far more complex. Owing to higher compression strains during the compression process, under the spherical compression, the compression strain at the maximal compression point in the contacting area can be located in first, second and third or all the three regions. As the compression strain at the maximal compression point exceeds critical compression strain at first region, the resultant forces withstood by both the spacer yarns in the contacting area should be separately calculated in different regions. When the compression strain at the maximal compression point is higher than critical compression strain at second region, the total resultant forces withstood by both the spacer yarns in the contacting area should be the sum of the forces integrated in regions I, II, and III. In this case, the intervals of the integration for first, second and third regions are obtained for the compression strain at any point on the upper surface of the fabric in the contacting area. By assuming a general region having modulus and intercept and an interval, a more general equation for each region of forces integrated in first, second and third regions can be derived by integrating concerned equation.

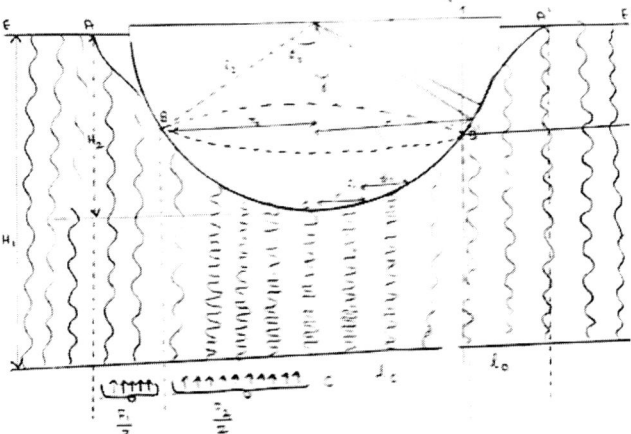

Figure 8 – Division of the contacting area by compression strains [29].

Hence, the calculation of the resultant force withstood by spacer filaments in the contacting area is to be carried out in three cases. In the first case, the compression strain at the maximal compression point is smaller than the first critical compression strain. In this case, the resultant force withstood by spacer filaments in the contacting area can directly be calculated with the help of equation by replacing some parameters. In the second case, the compression strain at the maximal compression point is between first and second critical compression strain. In the third case, the compression strain at the maximal compression point is bigger than the second critical compression strain..

6.9   Calculations vs measurements

In the case of all knit spacer fabrics the spherical compression curves obtained by both calculations and measurements have been depicted in figure 9.

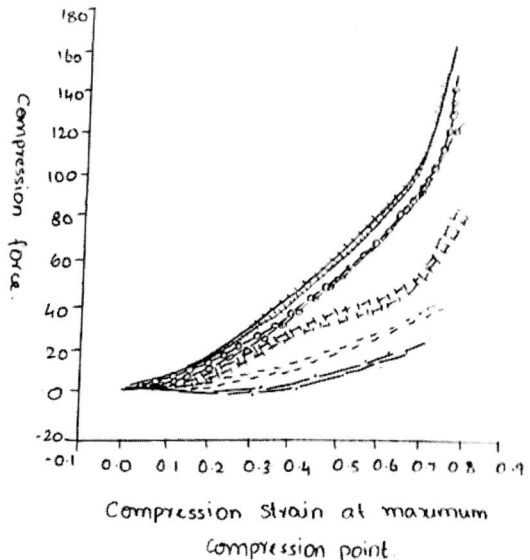

Figure 9 – Comparison between calculations and measurements [29].

The curves calculated are found to fit well with the experimental ones. Three compression parameters that can be used to characterize the compression behaviors of knitted spacer fabrics have been chosen so as to make better comparison. These are the maximal compression force and compression work at the compression strain of 0.70, and linear degree, which is defined as the ratio between the areas depicted in Figure 10. The values of these parameters from the model as well as the experiments have been determined. The maximal difference between the calculations and measurements is found to be less than 18.2%, which implies that the results from calculations and experiments agree well. Hence, the spherical compression of the knit spacer fabrics can be predicted from the plane compression test results using the theoretical model.

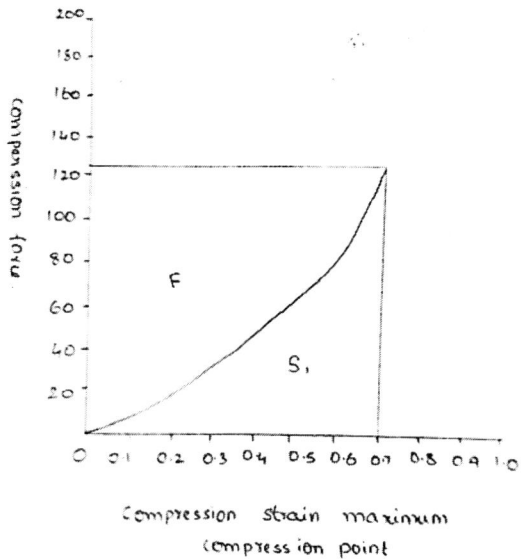

Figure 10 – Definition of the linear degree of a curve [29].

6.10 References

[1]   Liu YP and Hu H. Sound absorption behavior of knitted spacer fabrics. Textile Res J 2010; 80(18): 1949–1957.

[2]   Dias T, Monaragala R, Needham P and Lay E. Analysis of sound absorption of tuck spacer fabrics to reduce automotive noise. Measurement Sci Technol 2007; 18: 2657–2666.

[3]   Dias T, Monaragala R and Lay E. Analysis of thick spacer fabrics to reduce automobile interior noise. Measurement Sci Technol 2007; 18: 1979–1991.

[4]   Borhani S, Seirafianpour S, Ravandi SAH, Sheikhzadeh M and Mokhtari R. Computational and experimental investigation of moisture transport of spacer fabrics. J Eng Fibers Fabrics 2010; 5(3): 42–48.

[5]   Doyen W, Mues W, Molenberghs B and Cobben B. Spacer fabric supported flat-sheet membranes: a new era of flat-sheet membrane technology. Desalination 2010; 250: 1078–1082.

[6]   Bagherzadeh R, Montazer M, Latifi M, Sheikhzadeh M and Sattari M. Evaluation of comfort properties of polyester knitted spacer fabrics

finished with water repellent and antimicrobial agents. Fibers Polym 2007; 8(4): 386–392.

[7] Ye X, Fangueiro R, Hu H and Arau´ jo MD. Application of warp-knitted spacer fabrics in car seats. J Textile Inst 2007; 98(4): 337–344.

[8] Abounaim M, Hoffmann G, Diestel O and Cherif C. Development of flat knitted spacer fabrics for composites using hybrid yarns and investigation of two dimensional mechanical properties. Textile Res J 2009; 79(7): 596–610.

[9] Ye X, Hu H and Feng XW. Development of the warp knitted spacer fabrics for cushion applications. J Industrial Textiles 2008; 37(3): 213–223.

[10] Yip J and Ng SP. Study of three-dimensional spacer fabrics: physical and mechanical properties. J Mat Proc Technol 2008; 206: 359–364.

[11] Yip J and Ng SP. Study of three-dimensional spacer fabrics: molding properties for intimate apparel application. J Mat Proc Technol 2009; 209: 58–62.

[12] Liu YP and Hu H. Compression property and air permeability of weft knitted spacer fabrics. J Textile Inst 2011; 102(4): 366.

[13] Miao XH and Ge MQ. The compression behaviour of warp knitted spacer fabric. Fibers Textiles East Eur 2008; 16(1): 90–92.

[14] Mecit D and Roye A. Investigation of a testing method for compression behavior of spacer fabrics designed for concrete applications. Textile Res J 2009; 79(10): 867–875.

[15] Mecit D and Roye A. A study on the compression behavior of spacer fabrics designed for concrete applications. Fibers Polym 2009; 10(1): 116–123.

[16] Vassiliadis S, Kallivretaki A, Psilla N, Provatidis C, Mecit D and Roye A. Numerical modelling of the compression behaviour of warp-knitted spacer fabrics. Fibers Textiles East Eur 2009; 17(5): 56–61.

[17] Sheikhzadeh M, Ghane M, Eslamian Z and Pirzadeh E. A modeling study on the lateral compressive behavior of spacer fabrics. J Textile Inst 2010; 101(9): 795–800.

[18] Liu Y, Hu H, Zhao L and Long H. Compression behavior of warp-knitted spacer fabrics for cushioning applications. Textile Res J 2011; 82(1): 11–20.

[19] Hou X, Hu H and Liu Y. A study of compression mechanism of warp-knitted spacer fabric. Computers, Materials and Continua 2011; 23(2): 119–134.

[20] Shen Y, Liu H and Qian J. Spherical compression test and simulation analysis for warp-knitted spacer fabric. J Hunan Univ Technol 2008; 22(5): 1–4.

[21]   Puttock MJ and Thwaite EG. Elastic compression of spheres and cylinders at point and line contact. National Standards Laboratory Technical paper No. 25, Commonwealth Scientific and Industrial Research Organization. Melbourne: Australia, 1969.

[22]   Du Z and Hu H. A study of spherical compression properties of knitted spacer fabrics part I: theoretical analysis. Textile Res J 2012; 82(15): 1569–1578.

[23]   Ye X, Fangueiro R, Hu H, et al. Application of warp knitted spacer fabrics in car seats. J Text Inst 2007; 98: 337–344.

[24]   Miao XH and Ge MQ. The compression behavior of warp knitted spacer fabric. Fibres Text East Eur 2008; 16: 90–92.

[25]   Armakan DM and Roye A. A study on the compression behavior of spacer fabrics designed for concrete applications. Fiber Polym 2009; 10: 116–123.

[26]   Liu Y, Hu H, Zhao L, et al. Compression behavior of warp-knitted spacer fabrics for cushioning applications. Textile Res J 2011, 82(1): 11–20.

[27]   Hou X, Hu H and Liu Y. A study of compression mechanism of warp-knitted spacer fabric. Comput Mater Continua 2011; 23: 119–134.

[28]   Zhaoqun D and Hong H, A study of spherical compression properties of knitted spacer fabrics Part I: Theoretical analysis, Textile Research Journal, 82(15) 1569–1578

[29]   Zhaoqun D and Hong H, A study of spherical compression properties of knitted spacer fabrics Part II: Comparison with experiments, Textile Research Journal, 83(8)794.

Chapter 7

Analysis of the compressive behavior of warp knit spacer fabrics for evaluating suitability in cushioning applications

Summary

The compression behavior of warp knit spacer fabrics intended for cushioning applications have been studied by means of compression stress-strain curves as well as efficiency diagrams. Analysis has been done regarding the influences of various structural factors which comprise the spacer yarn inclination angle and fineness, fabric thickness, and outer layer structure. A number of findings emerged based on the experimental results and analysis. In considering cushioning applications, the warp knit spacer fabrics constitute a perfect class of energy absorbers. By mere change of their structural factors, their energy-absorption capacity can easily be tailored to suit specific areas of applications. The cushioning performance of the fabrics can be assessed effectively by the efficiency diagram. In order to choose suitable fabrics working at permitted stress levels at a known level of absorbed energy, the efficiency diagram can be effectively used. The cushioning performance and compression behavior are influenced by all the structural factors. The fabrics with lower spacer yarn inclination angle, higher fabric thickness, finer spacer yarns, and larger size mesh of the outer layers can be used to absorb lower energy with higher efficiency. Whereas, greater energy can be absorbed with greater efficiency by the fabrics having higher spacer yarn inclination angle, smaller fabric thickness, coarser spacer yarns, and smaller size mesh of the outer layers. Hence, choice of appropriate structural factors becomes crucial in order to design a spacer fabric having the necessary compression behavior.

7.1   Introduction

The kinetic energy of the impacting mass can be dissipated by using cushioning materials while maintaining the highest load (or acceleration) below some limit [1]. Under a relatively constant stress they normally absorb kinetic mechanical energy under compression actions which span over a wide range of displacement. In compression of such materials the works done equals the kinetic energies of a mass that might impact on them. A cushioning material is capable of absorbing most of the energy of the impacting mass when it is designed with an increased displacement and a suitable level of constant stress. Hence, as the protected object absorbs the impact energy under constant stress with an increased displacement, it would not have to endure a concentrated high-energy or high-load impact in the event of directly being impacted by a mass. Many materials and structures are available with the above mentioned feature that satisfy cushioning applications, and some typical examples include airbags, bubble films, rubberized fiber cushioning, and polymer-based foams. But, the inferior comfort property renders such materials and structures unsafe for human body protection even though they offer good cushioning properties and economy. Warp-knitted spacer fabrics are three-dimensional textile structures consisting of two separate outer fabric layers joined together but kept apart by spacer yarns, which are generally monofilaments [2]. Double needle bar high speed raschel machines are used to knit such fabrics, wherein the two layers of fabrics are knitted together by spacer yarns and offer possibilities of designing with many varied structures. Also, polyester multifilaments and monofilaments are widely used as the materials for spacer fabrics that are market products and are economical. Warp knit spacer fabrics owing to economy, high productivity, and varied structures render them highly suitable for a number of applications [3-6]. Specifically, excellent transversal compressibility coupled with high permeability renders such fabrics highly suited for multifunctional clothing and technical applications. The compression properties of warp-knitted spacer fabrics have been reasonably studied, and most of these have reported that the overall compression load-displacement relationship of these fabrics involve 3 main phases, i.e., linear elasticity, plastic plateau, and densification. This is the typical behavior required by a cushioning material in compression. The findings reveal relatively constant loads at the plastic plateau phase. The findings have revealed that warp-knitted spacer fabrics are a new optional materials for cushioning applications. But, as per the available literature the observed plateau stage is not significant, and also its zone too short [3-6]. It implies that the total energy absorbed in the plateau zone by these reported fabrics is inadequate to identify them as good cushioning materials. The chapter focuses on investigation of the compression behavior of warp-knitted spacer fabrics particularly designed for cushioning applications. A number of spacer fabrics have been warp knit in an effort to enlarge the plateau zone and reasonably control the load levels at the plateau stage. The various structural factors have been varied and include spacer yarn inclination angle and fineness, fabric thickness, and outer layer structure. The investigation has aimed to design a warp knit spacer fabric with satisfactory cushioning property.

7.2   Technical details

Double needle bar Raschel machine having 6 yarn guide bars has been used to knit the warp knit spacer fabrics. During knitting the binding of the structure has been done by polyester multifilament yarn to create the binding of the structure for upper as well as lower fabric layers. The polyester monofilament has been used as spacer yarns to connect the two outer layers together [8]. The outer fabric layers have been knitted using 4 different structures - locknit, chain plus inlay, rhombic mesh, and hexagonal mesh. Suitable chain notation for each of these structures has been given. Three different yarn guide bar lapping movements were used for the spacer yarns to connect two outer layers with different inclination angles. The chain notation for each movement has been indicated. Test spacer fabrics have been produced with regard to various outer layer structures, various lapping movements of the spacer yarn guide bars, and various fabric thicknesses. An extra spacer fabric has been made with finer spacer yarn. During knitting the wale stitch density of the spacer fabrics has been kept constant.

INSTRON instrument has been used to test the spacer fabrics as per prescribed standards. Compression stress strain curves have been obtained.

7.3   Typical compression stress-strain relationship and cushioning behavior

The compression stress-strain curve of a spacer fabric is depicted in figure 1. The process of compression comprises of 4 phases so as to enable the analysis of the compression behavior of the fabric. The first phase is the initial phase, the second is the elastic phase, the third is the plateu phase and the final is the densification phase based on the changes in the slope of the curve. During the first phase, there is a lower slope caused by the compression of the loose outer layers and their ineffective constraint for the monofilaments. Some slippage of the monofilaments takes place in the outer layers, since each loose multifilament stitch around a monofilament cannot tightly constrain the monofilament during this phase [8]. But, the entire compressed multifilament stitches are transformed to a fastened microstructure as the fabric is further compressed into the second phase. In this stage, the monofilaments buckle at a larger scale and they are better fastened by the multifilament stitches. Further, there is a rapid increase of the compression stress, or in other words a stiffer mechanical behavior of the fabric can be seen.

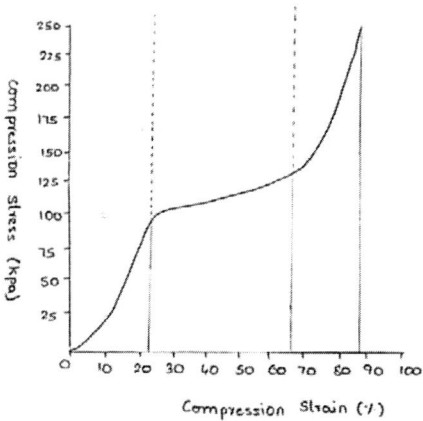

Figure 1 – Typical compression stress strain curve of a spacer fabric [8]

During the third phase an almost constant stress is achieved. At this phase the deformation mechanism of the fabric becomes very complex, and can be affected by the buckling, rotating, shearing, and intercontacting of the monofilaments and the contacting of the monofilaments with outer layers. The most significant influence factor for the fairly constant stress could be the inter-contacting of the monofilaments, of which the boundary conditions at their ends contacting with the outer layers are not constant. Because of the swift densification of the complete fabric, the compression during final phase exhibits a rapid increase in the stress. During this phase, the monofilaments within the fabric collapse and contact each other, and hence result in a really high stiffness. Since an almost constant compression stress is achieved in the third phase till a displacement that is above half the initial fabric thickness and corresponds to nearly 50% strain, the above analysis reveals that the spacer fabric possesses good cushioning effect. Such behavior is precisely what is expected of an ideal energy absorber. The energy absorbed by the fabric within the fairly constant stress is represented by the area under the curve from the first phase till the end of the third phase. The compression test has been carried out at a very low strain rate. Since the cushioning performance of warp knitted fabric is relatively equivalent to the absorbed kinetic energy of a mass that might impact on the fabric, the absorbed energy calculated at low strain rate can still be used as a good reference to optimize the cushioning performance eventhough the fabric will not have exactly the same behavior at higher strain rates under impact. A material absorbing optimum energy must be capable of dissipating the impact energy and at the same time maintain the load below the permitted limit. Hence, two criteria have to be considered. The first is the amount of the energy necessary to be absorbed by the impacted object, and the second is the stress with specific impact area to be permitted. During the densification phase low level of energy is absorbed by the fabric. However, there is a steep increase in the stress. Under this condition, for preventing the unpredictably increased reacting stresses in the densification phase, considering a specific application, it is preferable that the fabric dissipates all the energy before reaching the densification stage. Also, care needs to be taken to see that the stress in the plateau zone of the fabric is lower than the maximal stress permitted to the protected object. Hence, for a warp knitted spacer fabric to satisfy the needs of a particular application, optimization has to be done with regard to the amount of the energy absorbed before the densification stage and the stress level at the plateau stage.

7.4   Energy-absorption diagram

From the above study it can be seen that the determination of the energy absorbed by a spacer fabric under compression becomes crucial. It would be much more preferable to use the energy absorption diagram to understand better the cushioning behavior of the spacer fabric, despite the stress-strain curve being able to directly indicate the energy-absorption behavior of a spacer fabric. An energy-absorption diagram is obtained by plotting the absorbed energy per unit volume as a function of the stress. The area under the stress strain curve is the calculated absorbed energy/unit volume.

A cushioning material is expected to absorb the energy generated during impact, to the possible extent and transmit a force lower than the permissible limit. Hence, a cushioning material could be defined as one that transmits a permissible constant force to the protected object over a given range of compression strains [8]. But, this type of cushioning material does not exist and actual materials usually have the stress-strain curve conforming to that in Figure 2. For studying the energy absorption process the energy absorption efficiency can be used so as to better understand the energy-absorption capacity of a cushioning material. The efficiency is the ratio of the energy absorbed by a real cushioning material compressed to a given strain and energy absorbed by an ideal cushioning material that transmits a constant stress of the same value at the same given strain. The efficiency can be plotted as a function of the stress to obtain the indication for optimum usage, and proves useful. The efficiency can be expressed in the form of equation considering the area and the thickness of the cushioning material, its stress and strain.

The absorbed energy-stress as well as efficiency stress curves of fabric having locknit structure on both layers are illustratively depicted in Figure 3. The absorbed energy is found to steadily increase with the stress from the commencement of compression. But, there is a dramatic increase in the absorbed energy as the stress reaches towards the plateau stress despite the stress being maintained nearly constant in this zone. The absorbed energy then slowly increases with a rapid increase in the stress. The associated stress of a fabric for a given amount of energy to be absorbed is determined from the absorbed energy-stress curve, and hence it is convenient to choose a suitable fabric or to optimize the fabric performance for a particular application for which the amount of energy to be absorbed and the permissible stress level are predefined. The similar tendency is seen in the case of the efficiency-stress curve, the same tendency is noticed till the commencement of densification phase. At the end of the plateu phase the maximum energy-absorption efficiency is achieved. Beyond this point, there is rapid increase of the fabric density because of the densification of the structure. The point at the maximum energy-absorption efficiency can also be considered a critical point between the plateau zone and the densification zone. This point is used to define the plateau stress, which is calculated by dividing the plateau load to the area of the specimen. The contours of the energy absorbed/unit volume(iso energetic curves) is calculated so as to better understand the relationship between the energy-absorption efficiency and the stress (Figure 2). The contours show that for a given stress the greater is the absorbed energy as the efficiency-stress curve becomes greater. On the other hand, the efficiency-stress curve can intuitively show the maximum efficiency point, the contours of the energy absorbed per unit volume can clearly demonstrate the amount of the energy at this point. Also, the efficiency-stress curves as well as contours of the energy absorbed can be prove highly beneficial for the determination of the preferable working range of the fabric.

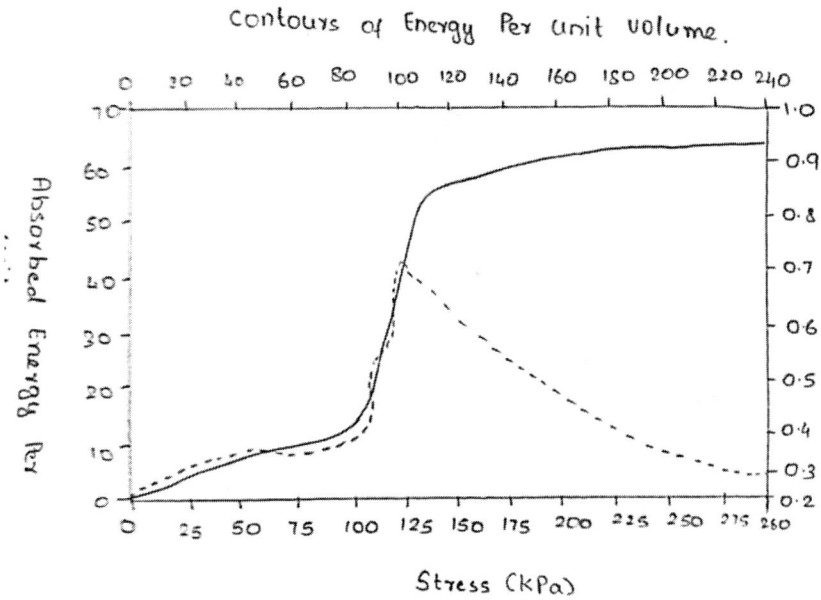

Figure 2 – Typical energy stress and efficiency curves of a spacer fabric[8]

Hence, at a given amount of absorbed energy, the better cushioning performance for a spacer fabric should function at a lower stress level but at a greater efficiency. Since the efficiency diagram comprising of efficiency-stress curves as well as the contours of the energy absorbed is user friendly and provides a strong base, it has been used to study the influences of various structural factors that consist of the spacer yarn inclination angle and fineness, fabric thickness, and outer layer structure, on the compression behavior of the warp-knitted spacer fabrics for cushioning applications.

7.5   Influence of the inclination of the spacer yarn

The two surface layers of the spacer fabric are connected by the spacer yarns. The number of underlapped needles between front and back needle bars determine the inclination angle and length of the spacer yarns. Three fabrics having various numbers of underlapped needles have used for comparison so as to study the influences of the inclination angle on the compressive behavior of the spacer fabrics. The fabrics possess similar outer layer structure. Simultaneously, their other parameters remain almost similar. On the other hand the angles formed between the outer layers and the left oblique spacer yarns and the angles between the outer layers and the right oblique spacer yarns are nearly the same for similar fabrics [8]. Such differences arise from the problem in producing the fabrics with exactly symmetrical spacer yarns in a complex manufacturing and finishing process. Apparently, with the increase in the number of underlapped needles there is reduction in the spacer yarn inclination angle. The greatest difference of the spacer yarn inclination angles between the left and right oblique spacer yarns has been noticed in the case of spacer fabric having chain and inlay structure with single lapping of spacer yarns. Such high difference in spacer yarn inclination angles renders the fabric(chain plus inlay structure and single lapping of spacer yarns) asymmetric and unstable in the fabric structure that results in the shearing between two outer layers along the coursewise direction under compression. Hence a marked drop in the plateau zone is noticed in the compression stress-strain curve of the fabric (Figure 3(a). The outer layer of the fabric surface has been stuck to the compression plates before testing so as to avoid the influence of shearing. Figure 3(a) also depicts the test result under such a state. The marked drop vanishes as expected. Figure 3(a) depicts the curves for fabrics having chain and inlay structure with two and three lappings of spacer yarns for the comparison.

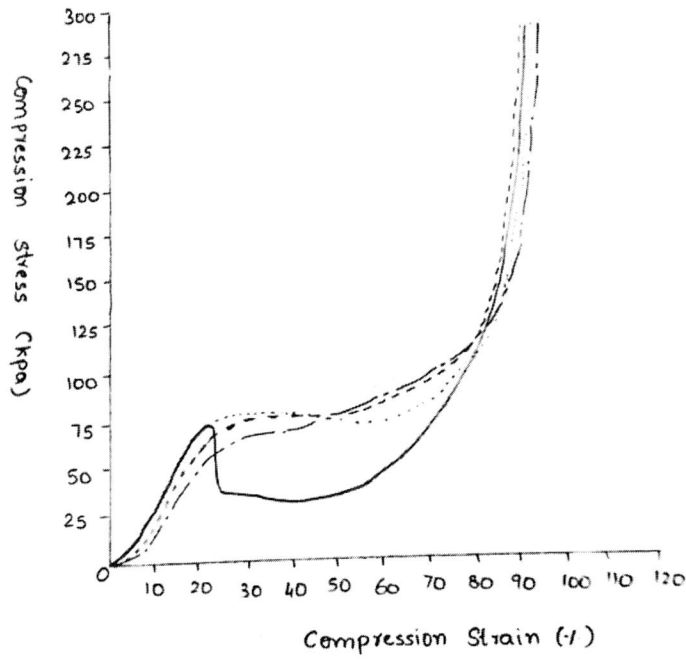

(a)

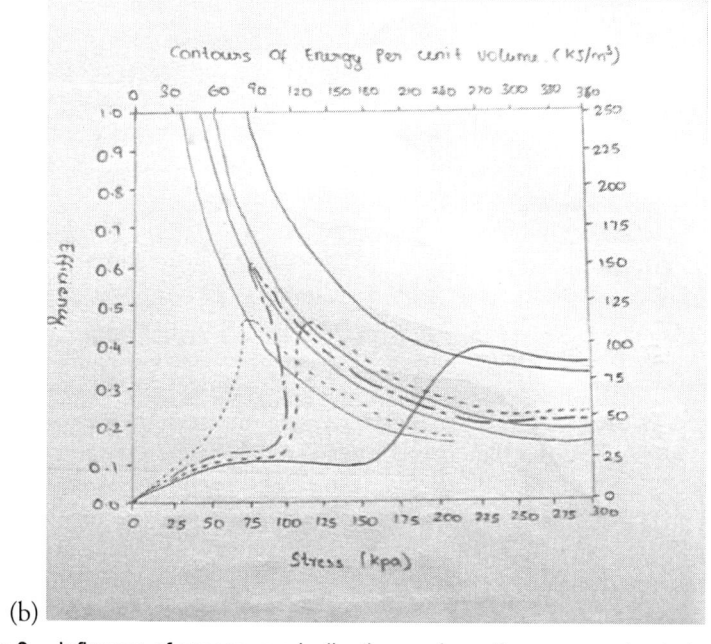

Figure 3 – Influence of spacer yarn inclination angle on the compression behavior of spacer Fabrics [8]

a) Stress-strain curves

b) Efficiency diagram

With the decrease in the spacer yarn inclination angle there is reduction in the compression resistance of the spacer fabrics at the initial and elastic stages. But, the condition reverses after the strain reaches about 47.5%. The drop in the stress is evident from this point to the end of the plateau zone in the case of the fabric with chain and inlay structure and single lapping of spacer yarn. However, in the case of the fabric with chain and inlay structure and three lappings of spacer yarn there is a slight increase. In the case of the fabric with chain and inlay structure and two lappings of spacer yarns, the stress remains almost constant. Figure 3(b) depicts the efficiency diagram, i.e., efficiency-stress curves and contours of the energy absorbed/unit volume is shown. All the three fabrics show same maximum efficiencies which is less than 70%, and reduce with decreasing spacer yarn inclination angle. With the increase in the spacer yarn inclination angle, the stress at the maximum efficiency point increases. At lower energy level, the fabric having chain and inlay structure with 3 lappings of spacer yarns show the highest efficiency and hence has the least stress for a given energy. But, its efficiency reverses beyond this energy level. The fabric having chain and inlay structure with a single lapping of spacer yarn behaves opposite to that of fabric with chain and inlay structure and having 3 lappings of spacer yarns. Fabric having chain and inlay structure and with two spacer layers shows moderate performance in all situations. In Figure 3(b), the contour of energy passing by the maximum efficiency point is also demonstrated for each fabric. The fabric having chain and inlay structure with two spacer layers can absorb more energy than others over a wide level of energy, and hence exhibits better cushioning performance. Briefly stated, fabric having smaller spacer yarn inclination angle exhibits better cushioning performance at low energy level and fabric having bigger spacer yarn inclination angle shows preferable cushioning property at high energy level.

7.6   Influence of the fabric thickness

In order to study the influence of the fabric thickness on the compression behavior of the warp knit spacer fabrics, 4 types fabrics have been knitted having equal number of needles underlapped for the spacer yarns and the same outer layer structure but with varying thicknesses. Figure 4(a) depicts the stress-strain curves of the fabrics. The compression resistance is found to decrease with the increase in the fabric thickness. A longer and lower the plateu zone is noticed in the case of a thicker fabric. Figure 4(b) depicts the efficiency diagram. Owing to its low plateu a thicker fabric can absorb a defined amount of energy in a larger deformation but at a lower stress level. On the contrary, the thinner fabric absorbs the same amount of energy in a lower deformation but at a greater stress level. Also, the thicker fabric attains its maximum efficiency point at a lower stress and energy level. On the other hand the thinner fabric attains its maximum efficiency point at much greater stress and energy level. Under such conditions, the fabrics having various thicknesses find various ranges of applications and cannot be directly compared [8]. The fabric thickness has to be chosen based on the amount of the energy to be absorbed and the permitted level of stress. Hence, the efficiency of a fabric can be optimized for a particular application by use of the efficiency diagram.

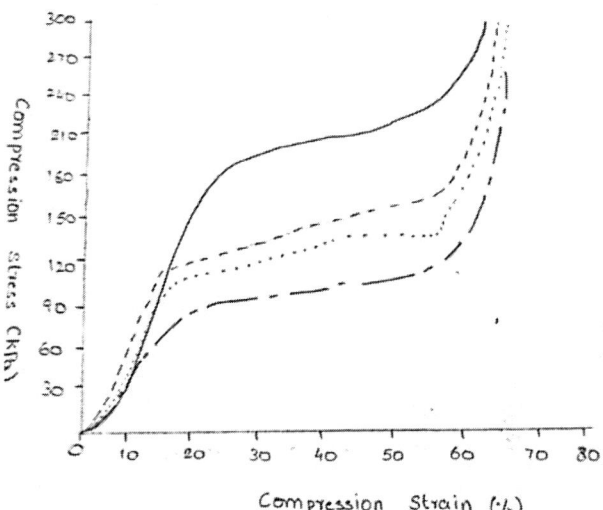

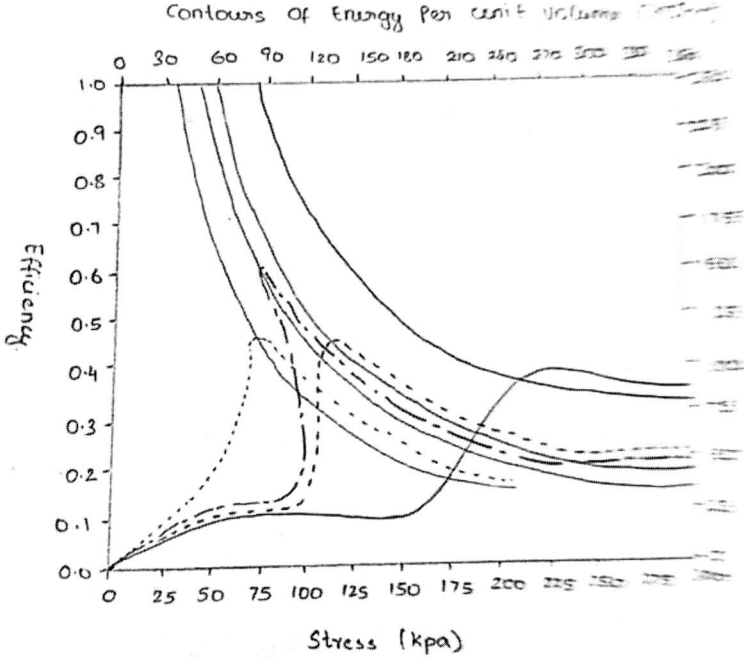

Figure 4 – Influence of fabric thickness on the compression behavior of the spacer fabrics [8]

a)   Stress-strain curves
b)   Efficiency diagram

7.7   Influence of the fineness of spacer yarn

In order to study the influence of the fineness of spacer yarns on the compression behavior, two fabrics having the equal number of needles underlapped for the spacer yarns and the similar outer layer structure but having two different diameters of spacer yarns have been selected. The two fabrics have nearly equal thickness, and also have very close stitch densities in their outer layer. Figure 5(a) and (b) depicts the compression stress-strain curves of the two fabrics. The fabric produced from coarser spacer yarn exhibits greater compression resistance and a greater value plateau [8]. The maximum efficiency can be achieved at lower stress and energy for fabric produced from finer spacer yarns owing to its lower plateau level. But, in the case of fabric made from coarser spacer yarn the energy at the maximum efficiency is far greater in comparison with that of fabric produced from finer spacer yarn, which shows that fabric produced from coarser spacer yarn can absorb more energy but at a greater level of stress. It can be emphasized that fabric produced from the spacer yarn of diameter of 0.16mm has a high efficiency in a range of stresses lower than 100 kPa. Hence, for a given absorbed energy, the associated stress of the spacer fabric can be varied by simply adjusting the diameter of the spacer yarns based on the maximum stress permissible for a protected object to attain a high efficiency of energy absorption. Therefore, fabric produced from finer spacer yarn is suitable for lower energy absorption and lower stress level and fabric produced from coarser spacer yarn is suitable for greater energy absorption and stress level.

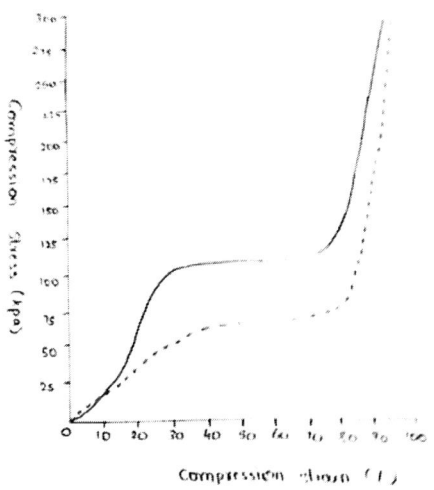

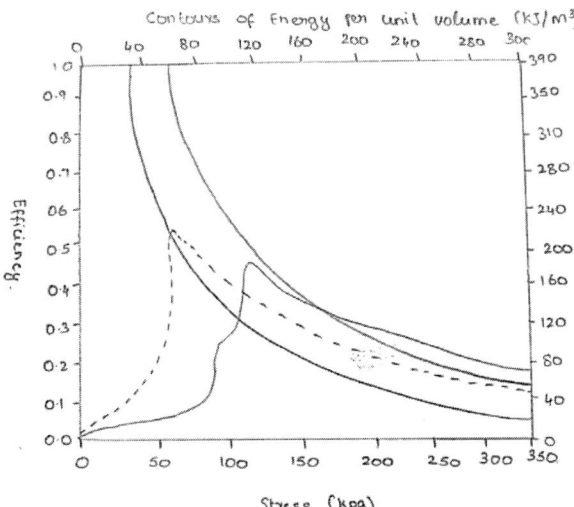

Figure 5 – Influence of spacer yarn fineness on the compression behavior of spacer fabrics [8]

a)   Stress-strain curves
b)   Efficiency diagram

7.8   Influence of structure of the outer layer

The structure of the outer layer affects the distribution, binding condition, and inclination angle of the spacer yarns since the monofilaments in the spacer layer are bound by the multifilament stitches in the outer layer. In order to study the influence of the outer layer structure on the compression behavior of the spacer fabric, six fabrics having equal number of underlapped needles for spacer yarns and almost equal thickness but having various outer layer structures have been selected [8]. The outer layer structures can be divided into the following categories,

   a)   Both outer layers having close structure

   b)   The top layer having open structure and the bottom layer having close structure

   c)   Both outer layers having open structure.

The close structures comprise locknit and chain plus inlay, and the open structures comprise rhombic mesh and hexagonal mesh. The distribution, shape and size of the stitche in the outer layers can be clearly noticed from the outer layers with such structures. The outer layer structures could slightly affect the stitch density of the outer layers and the spacer yarn inclination angle, though such parameters are maintained constant during knitting. Figure 6(a) depicts the compression stress-strain curves. During the initial and elastic stages, both fabrics (with hexagonal mesh structure in top layer and chain plus inlay structure in bottom layer for one fabric and hexagonal mesh structures for both layers in another fabric) show the lowest compression resistance among the fabrics studied [8]. At the plateau stage, the fabric having hexagonal mesh structure in both layers and double spacer layer has the lowest value, but fabric Hexagonal mesh structure in top layer and chain plus inlay in bottom layer with double spacer layer has close value to that of fabric with locknit structure in both layers and double spacer layers and fabric having chain plus inlay and single spacer layer with close outer layer structures. Such differences in compression behavior arise due to the uneven distribution of the stitches, the changes in stitch density of the outer layers and the spacer yarn inclination angle because of the changing outer layer structure. The fabrics having rhombic mesh structures in both layers and double spacer layers and fabric having rhombic mesh structure in top layer and chain plus inlay structure in bottom layer with double spacer layer have greater value plateau compared with other fabrics. It indicates that rhombic mesh structure has better stability and is more appropriate for absorbing greater energy. The findings also reveal that the fabrics having close outer layer structures exhibit moderate compression resistance and plateau values. Figure 6(b) depicts the efficiency diagram. The fabrics having open structures in both of outer layers (fabric with rhombic mesh structure in both layers and fabric with hexagonal mesh in both layers and double spacer layers) exhibit the lowest and highest values, respectively, of the energy absorbed at their maximum efficiency point, the fabrics with close structures in one or two outer layers (Fabric with locknit structure in both layers and double spacer layers, fabric with chain plus inlay structure in both layers and double spacer layers, fabric with rhombic mesh structure in top layer and chain plus inlay structure in bottom layers and double spacer layers, and fabric with hexagonal mesh structure in top layer and chain plus inlay structure in bottom layer and with double spacer layers) have middle values between the lowest and highest values. Such findings reveal that a large range of the variations in the energy absorbed for various applications can be achieved with the open structures. On the other hand the fabric having the hexagonal mesh structure in both outer layers can be used to absorb the energy at lower stress level, the fabric with the rhombic mesh structure can be used to absorb the energy at higher stress level. Hence, the variation of the outer layer structure can be another approach for choice of the fabrics, which can absorb the same quantity of the energy, but with different stress levels for varied end uses.

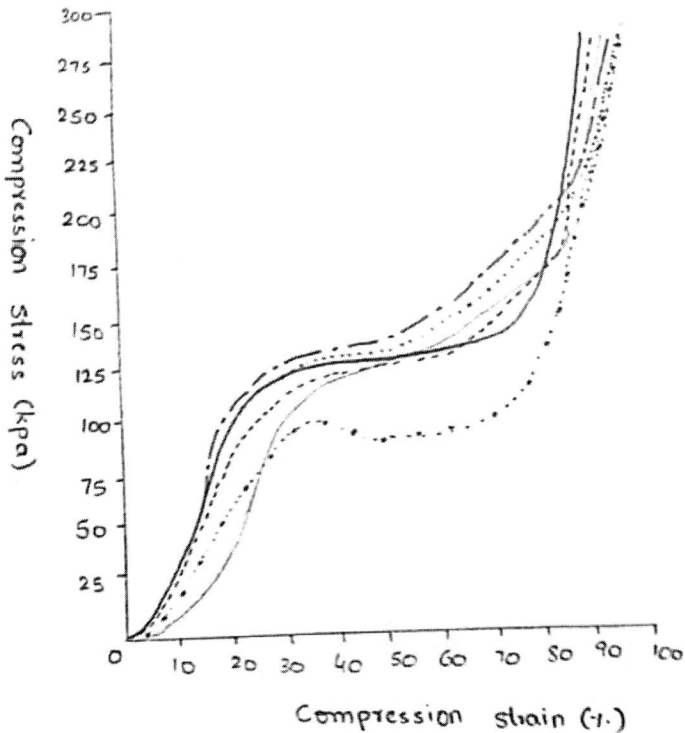

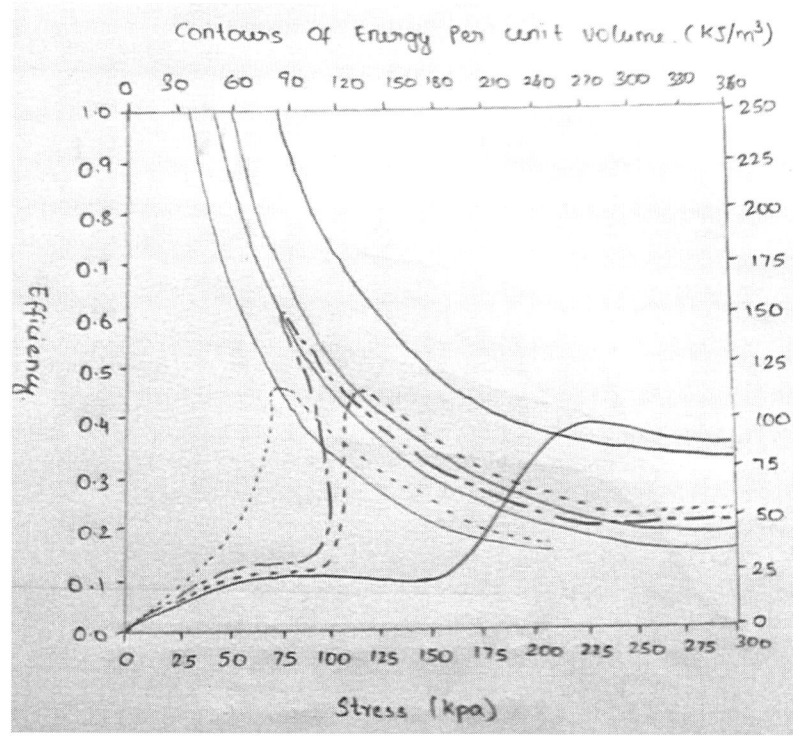

Figure 6 – Influence of the outer layer structure on the compression behavior of the spacer fabrics [8]

d)   Stress-strain curves

e)   Efficiency diagram

7.9   References

[1]   Avalle M, Belingardi G and Montanini R. Characterization of polymeric structural foams Under compressive impact loading by means of energy-absorption diagram. Int J Impact Eng 2001; 25: 455–472.

[2]   Liu YP and Hu H. Compression property and air permeability of weft-knitted spacer fabrics. J Text Inst 2011; 102: 366–372.

[3]   Ye X, Fangueiro R, Hu H and Arau´ jo M. Application of warp-knitted spacer fabrics in car seats. J Text Inst 2007; 98: 337–344.

[4]   Miao XH and Ge MQ. The compression behavior of warp knitted spacer fabric. Fibres Text East Eur 2008; 16: 90–92.

[5]   Armakan DM and Roye A. A study on the compression behavior of spacer fabrics designed for concrete applications. Fiber Polym 2009; 10: 116–123.

[6]   Mecit D and Roye A. Investigation of a testing method for compression behavior of spacer fabrics designed for concrete applications. Text Res J 2009; 79: 867–875.

[7]   Miltz J and Gruenbaum G. Evaluation of cushioning properties of plastic foams from compressive measurements. Polym Eng Sci 1981; 21: 1010–1014.

[8]   Yanping L, Hong H, Li Z and Hairu L, Compression behavior of warp-knitted spacer fabrics for cushioning applications, Textile Research Journal 82(1) 11–20.

Chapter 8

Impact and compression after impact properties of warp knitted spacer fabrics

Summary

In the case of warp knit spacer fabrics, the impact behavior and damage characteristic after impacts have been studied. Spacer fabrics having various structures have been assessed for the two properties by using parameters such as peak force, absorbed energy, damage depth, and drop-off rate of residual strength. More absorbed energy, lower peak force, lower damage depth, and a lower drop off rate of residual strength have been considered for a spacer fabric having good protection performance. A number of interesting findings have emerged based on the experimental results and analysis. During an impact process the variation of force is due to the combined influence of two factors, namely, displacement and failure of spacer yarns in the center zone. Also, as the displacement increases the force increases and reduces due to failure of the spacer yarns. In the case of fabrics having the same needle bar distance (different surface structures and finenesses of spacer yarns), fabrics having coarser spacer yarns and close surface structure show many beneficial outcomes, like lower peak force, larger absorbed energy, and lower depth value, but higher drop-off rate of residual strength. While considering the specific impact method (flat-sphere), fabric with a mesh structure is body-fitted, for meshes can open and shut more freely according to the shape of wrapped objects. But, mesh sides are inclined to collapse on surface layers due to much bigger size of meshes. This implies that spacer yarns cannot be protected perfectly. In the case of fabrics having various needle bar distances (same surface structure and fineness of spacer yarns), fabric having higher thickness performs perfectly in the experiments (lower peak force, more absorbed energy, lower damage depth, and lower drop-off rate), but a very thick protector makes a person's body uncomfortable. It is necessary to consider the balance between the protective performance and comfort by choice of an appropriate fabric thickness for the particular protective application. As the energy levels increase, the peak force, absorbed energy, and damage depth increase, whereas residual strengths approximate to each other, in the case of a fabric subjected to various energies.

8.1   Introduction

Warp-knitted spacer fabrics are basically 3D structures that have two layers of fabric bound together by spacer yarns which keep the layers apart [1]. Compared with Polyurethane foam, they possess much better moisture transmission characteristics, better pressure relief properties, higher air permeability, and lower heat resistance, when considered as a cushion material [2,3]. Hence, they have been widely used automobile textiles as cushions or car seats, sports textiles, and foundation garments like bra cups, pads for swimwear, and so on [3-7]. A number of investigations have been carried out on warp-knitted spacer fabrics that cover areas including on their static compression properties, sound absorption behavior, pressure distribution, air permeability, heat resistance, and so on [7-9]. There are a number of interesting studies relating to static compression behaviors of warp knitted spacer fabrics. The influence of some factors, such as material, pattern, threading, location angle and the number of spacer yarns on the compression behaviors have been studied [10].The lateral compressive behaviors of spacer fabrics have been investigated based on Van Wyk's equation [11]. Another area of study has focused on the indentation force deflection (IFD) properties of different warp-knitted spacer fabrics [12]. The mechanical and the stress– strain model of single spacer yarn of warp-knitted spacer fabric have been analyzed [13,14]. The energy efficiency and non-linear compression behavior of spacer fabrics have also been studied [15]. A number of research workers have studied the static compression properties of warp-knitted spacer fabrics as highlighted above. But, only few researches have emphasized on the impact behaviors of spacer fabrics. The impact method has been studied using a flat sphere. Compared with the impact method (flat flat) of earlier investigation, it is quite different [16]. For reducing fractures during motion, particularly in the case of aged people, spacer fabric as cushioning material is expected to be sewn into some parts of clothing corresponding to injury-prone joints of the human body, like the elbow, femoral greater trochanter, knee, etc [17]. A simplified model of the event was that of spacer fabric wrapped on a hemispherical steel anvil (joints of human body) impacted with a flat striker (ground).

8.2   Technical details

Raschel warp knitting machine has been used to produce warp-knitted spacer fabrics used in the investigation. The spacer fabrics have two kinds of surface structures of which the first one is closed structure without meshes and the second one is open structures with either small or large-size meshes based on their various shapes. Figures 1(a) to (c) depict pillar+weft insertion, rhombic mesh, and hexagonal mesh used for the surface layers, and their typical representatives respectively. The figures clearly depict that the fabric surface is plain in the case of pillar+weft insertion, whereas fabric surface layers having rhombic mesh have smaller meshes and fabrics with hexagonal mesh have larger meshes on their surface layers. The chain notations and materials for these samples are shown in Table 1. Fabrics of three different fabric thicknesses have been considered produced with different needle distances. Also, two different diameters of spacer yarns have been used. Fabrics have been heat set at 180oC for 3 minutes [18].

Special type of instron tester has been used to conduct the impact tests as per specific British standards relating to motorcyclists protective clothing against mechanical impact. The drop hammer falls freely and is automatically arrested by a secondary impact preventer after rebounding to keep it from a second strike. The striker face consists of polished steel having specified dimensions. The drop hammer has a mass of about 6.3 kgs. The warp-knitted spacer fabric has been wrapped on the hemispherical steel anvil and the four sides of the fabric are held by clamps on the plane support. Four levels of incident impact energies have been selected for the study. Three samples have been tested for each type of spacer fabric and every sample has been impacted once. In each type of fabric average value of three tests have been considered. The fabrics have been conditioned in standard atmosphere. After positioning a sample on the steel anvil well, the test was completed within 3 min. A special type of 3D laser scanner has been used to study damage deformations and depths of all the fabrics after impact. The distance between the fabric and transmitter is calculated by collecting the reflection sent from a laser by a transmitter. Hence, it is possible to get the height of each point on the surface of fabric. The laser beam moves along X and Y axis by a given scanning area guided by the transmitter. The interval of the X-axis was 0.001mm and that of the Y-axis was 0.1 mm. Compression-after-impact tests have been carried out on a precision instrument and the compression curves have been obtained from all the fabrics after conditioning in standard atmosphere. One way analysis of variance have been used to evaluate the values of peak force, absorbed energy, and damaged depth of the fabric samples in the tests for each dependent variable.

8.3   Impact behaviors of spacer fabrics

Figure 1 depicts the deformations that occur in fabric in the event of an impact. The figure (a) shows the side elevation of original fabric. The upper and lower surfaces of the fabric are plane and all the spacer yarns are slightly bent in their natural condition. There are tensile and compressive stresses produced on the upper and lower surfaces when the fabric is wrapped around a steel anvil as shown at b. The spacer fabric absorbs and stores energy given by the striker under the loading condition of the impact process as shown at figure c. On the other hand the fabric undergoes plastic and elastic deformations in order to decrease the force transmitted to the steel anvil [18]. Because of the specific contact method (flat sphere) the stress in the contact area is non-uniform. Specifically, there is maximum stress at the center of the contact zone and spacer yarns in the middle zone are compressed more seriously compared with those in neighbouring zone. As shown at figure during the unloading process the striker is rebounded and elastic energy stored in the spacer fabric is released, the spacer fabric recovers to some extent as well. There is formation of an almost circular pit after removal of the test fabric from the steel anvil as shown at figure e, under the compressive and tensile stresses on the upper and lower surface. Also, the spacer yarns are subjected to irrecoverable deformation in the contact zone as depicted at figure f.

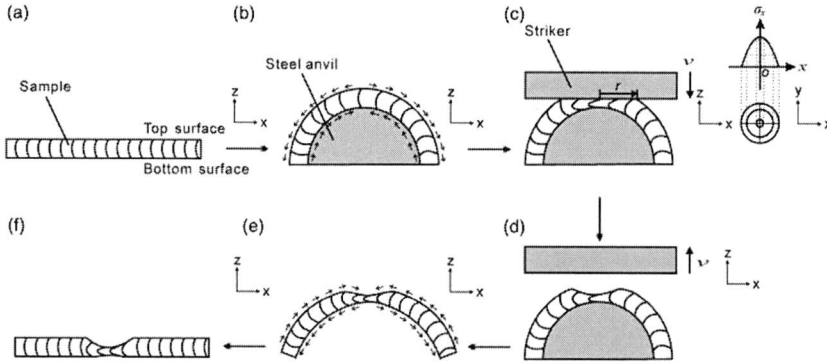

Figure 1 – Fabric deformation during impact process[18]

The force-time and force-displacement curves depicted in Figures 2(a) and (b) enable to study typical curves of test fabrics during impact tests. The figures show when the incident energy levels are increased in magnitude as well as configuration there is variation in the curves. There is increase in the highest displacement at peak forces. On the other hand there is reduction in contact time as the impact energy rises in the loading stage. During the phase of unloading the forces fall markedly and to a certain extent the displacements fall. The force values fluctuate around zero and finally move to their value along with vibrations of the machine. Also, between the drop hammer and secondary impact preventer there are vibrations. As the drop hammer and preventer are rigid bodies, the former cannot stop immediately after rebounding and being arrested by the latter. But as the vibrations or collisions decrease, it will stop gradually. There are two factors that affect change of force. One is the variation of displacement and the other is failure of spacer yarns in the central zone. As the displacement increases the compression strength increases [2]. Also, the force values are affected by the change of contact area arising from that of displacement (Figure 3).

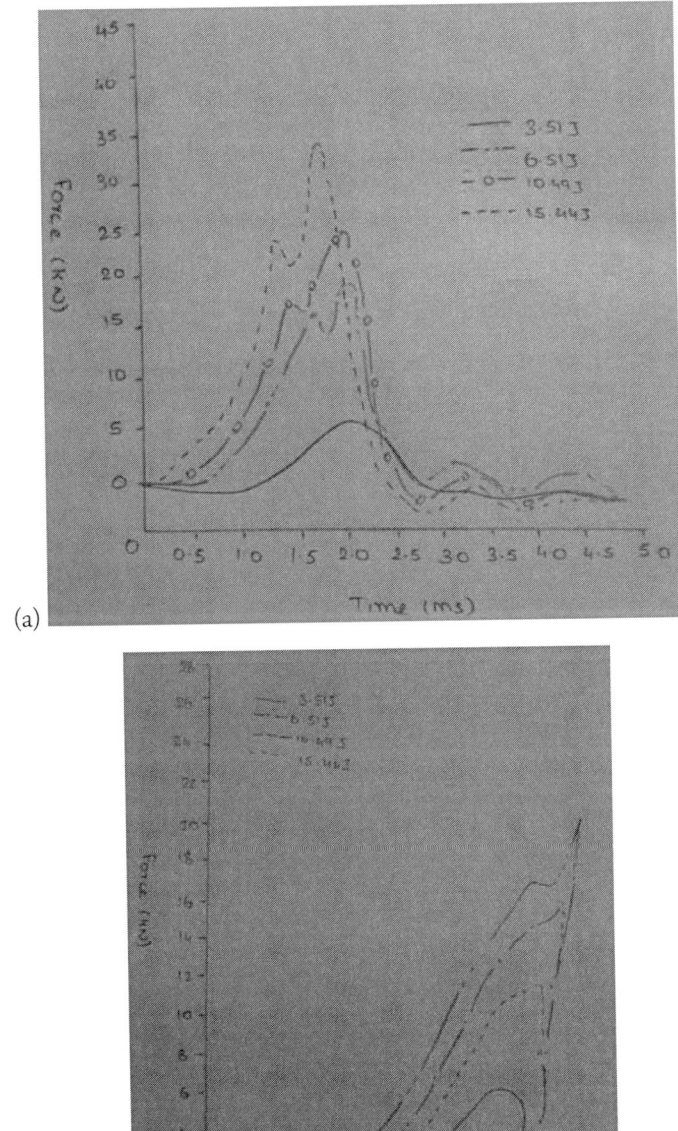

(a)

(b)

Figure 2 – Typical curves depicting force-time/displacement[18]

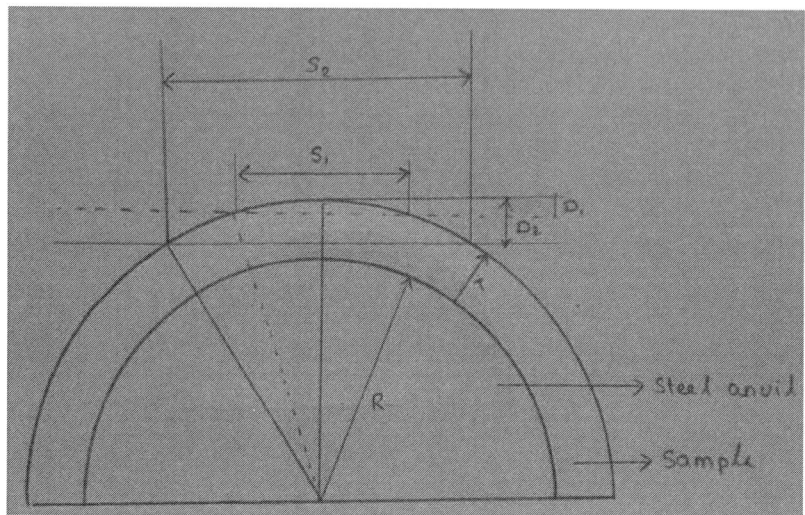

Figure 3 – Variation of contact area with displacement[18]

It has been found that as displacement increases the contact area increases. There is an increasing force on the striker as more spacer yarns exert force on the striker. There is failure of spacer yarns in the middle area. The spacer yarns at the middle are damaged more than others because of the non-uniform stress on the contact zone. The declination of load-bearing capacities of these damaged spacer yarns leads to the decrease of the force. Normal spacer yarns and damaged spacer yarns impacted under lower and higher energy have been compared. The total force acting on the striker by the above two factors and the change of force is based on which factor plays a major role. Due to displacement increase and relatively slight failure of spacer yarns there is a continuous increase of force in the loading stage for fabric subjected to lower impact energy. When impacted under higher energy the damage on the spacer yarns are more severe. As the test fabric is subjected to higher energy the impact force drops at some instants, owing to the failure of spacer yarns due to the change of contact area during loadings.

There are two types of energies that arise from the impact energy as the striker impacts the test fabric surface, namely rebounded energy/elastic energy that is stored elastically in the specimen and transferred back to the striker, and absorbed energy that is the sum of absorbed energy in the specimen by forming irrecoverable deformations, and the energy absorbed by the impact system in vibration, heat, inelastic behavior of striker, and supports.

It has been assumed that the energy absorbed by the impact system is negligible. Hence, the total energy is the product of rebounded energy and absorbed energy by specimen. Figure 4 also specifically depicts the energy-absorption rates (Eabs/Etotal) of fabric subjected to various energies.As the impact energies increase the ratio reduces, and some percentage of the impact energy can still be absorbed by the undamaged material for all energy levels. Figure 5 depicts the force displacement curves for all test spacer fabrics subjected to various impact energies with the objective of analyzing the influences of fabric structure, thickness, and yarn diameter on the impact behavior. The figure shows that change of forces for each test spacer fabric impacted under various energies has a tendency of that force's drop at some instants in the loading stage in higher energies. When considering various fabrics under equal impact energy, an ideal spacer fabric can absorb more energy, and thus has lower peak force. Figures 6(a) and (b) show the values of peak force and absorbed energy for all test spacer fabrics impacted under various energies. The results of significance test have been determined. Spacer fabrics have been knitted with the same needle bar distance and same fineness of spacer yarns, but with different surface structures, which are pillar+weft insertion, rhombic mesh, and hexagonal mesh, respectively. In the case of fabric having large size meshes the value of Fmax on surface layers is higher when compared with the small-size ones (fig 6(a)). Hence the former can be used to absorb less energy (Figure 6(b)). There are two main factors that attribute for this phenomenon. Firstly the fabrics having greater spacer yarn inclination angles can be used to absorb more energy in compression behaviors. Secondly, in the case of spacer yarns of fabric with large-size meshes are more inclined to collapse and fail to deform normally to absorb impact energies. Also, as parts of the spacer yarns are exposed in the open air and are not properly covered by surface layers, they are more prone to damage. Spacer fabrics have been constructed having the same surface structure and same fineness of spacer yarns, but different needle bar distances. In the case of fabric having higher thickness the value of Fmax is lower than that with the smaller one (Figure 6(a)). Hence the fabric with higher thickness can be used to absorb more total energy than the one with lower thickness (Figure 6(b)). Spacer fabrics have been constructed with the same surface structure and same needle bar distance, but different finenesses of spacer yarns. In the case of fabric having coarser spacer yarns the value of Fmax is lower than that having the finer ones (Figure 6(a)). Thus the fabric with coarser spacer yarns can be used to absorb more energy than the one with finer spacer yarns (Figure 6(b)). It has been found that there is a high significance on the influences of surface structure, fabric thickness, and spacer yarns' fineness of fabrics impacted under different energies on peak force and absorbed energy.

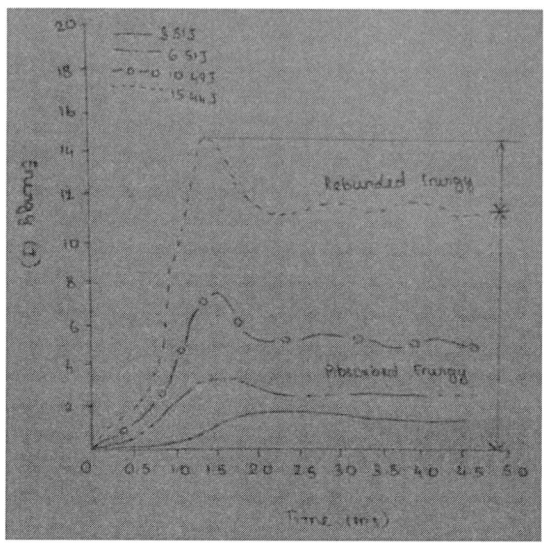

Figure 4 – Typical curves depicting energy-time[18]

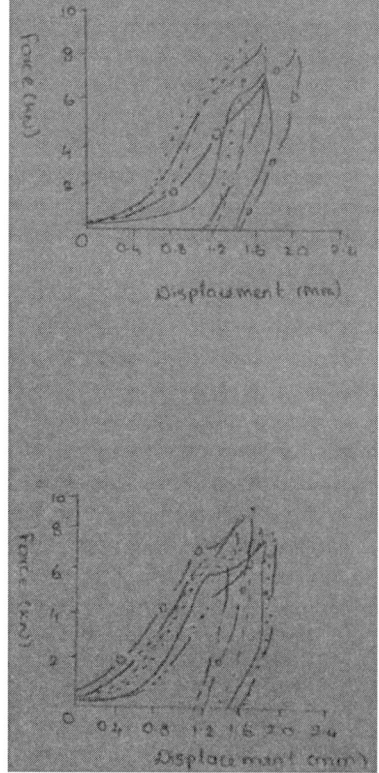

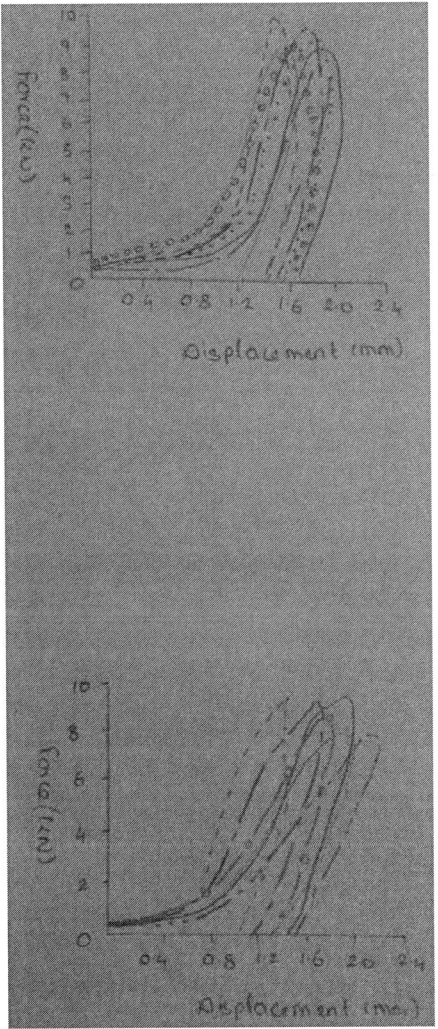

Figure 5 – Curves depicting force displacement for all samples[18]

8.4   Depth of damage and deformation

Despite the flat sphere technique (flat-sphere) being used for all impacts, it has been observed that there is difference in damaged surfaces of the test fabrics and their depth values. But, the ability of a fabric to offer resistance to impact are indicated by deformation and depth. Under the same conditions good impact resistance is indicated by small damage depth [18]. The damaged surface deformations of the test fabrics have been studied using 3D scanner. It has been observed that the surface deformations of fabrics with the same surface layers under four different impact energies resemble each other in shape but not in dimensions. The deformations of fabrics having various surface structures impacted under the same energy differ from each other. A number of height lines are also observed around the damaged areas for fabrics having rhombic and hexagonal mesh structure. As the surface layers of fabric having pillar and weft insertion structure(7.38 mm thickness) are plain and tight, while those of fabrics having rhombic and hexagonal mesh structures have many meshes on the outer layers, laser beams derived from the transmitter could irradiate the closed top surface of fabric with pillar and weft insertion structure(7.38 mm thickness), while parts of them irradiated directly to the inner or even the bottom of the fabrics having rhombic and hexagonal mesh structures because of the open structures on outer layers. The damaged deformation of fabric having pillar and weft insertion structure shows that the damaged zone tends to become localized and is more or less a round pit in shape. The deepest point occurs in the center of the pit. The damaged area appearances of fabric with pillar and weft insertion structure and rhombic mesh structure are similar (a round pit). But, in the case of fabric having hexagonal mesh structure, the hexagonal sides have been broken. Photographs of damaged fabrics show this more clearly. The photographs have revealed that the outer layers of fabrics having pillar and weft insertion structure and fabric having rhombic mesh structure are still fine, while the appearance of fabric having hexagonal mesh structure reveals the breakage of hexagonal sides and collapse of spacer yarns around them, as greater stress concentrates on the mesh sides and parts of spacer yarns are exposed in the open air. As per the laser scanner damage depths for all the fabrics can be calculated, and the results have been determined. Depth values of various specimens impacted under the same energy and those of the same fabric with pillar and weft insertion (7.38 mm thickness) subjected to different impact energies have been determined. In the case of fabric having pillar and weft insertion structure (7.38 mm thickness), fabrics with rhombic and hexagonal mesh structures, the depth of fabric having pillar and weft insterion structure (7.38 mm thickness) is lesser in comparison with that of fabric having rhombic mesh structure. It implies that the impact resistance of fabric having a closed surface layer (fabric having pillar and weft insterion structure (7.38 mm thickness)) is better compared with small-size meshes on surface layers (Fabric having rhombic mesh structure). On the one hand the damage depth of fabric having large-size meshes on the surface layers is not compatible with those of fabrics having pillar and weft insertion structure (7.38 mm thickness) and rhombic mesh structure, for their damage conditions are different, the surface yarns of fabrics having hexagonal mesh structure are broken while those having pillar and weft insertion structure (7.38 mm thickness) and rhombic mesh structure are not. Nearly the entire absorbed energy is transformed into irrecoverable deformation energy of spacer yarns. Hence, a pit is created on the fabric surfaces for fabrics having pillar and weft insterion structure (7.38

mm thickness) and rhombic structure. On the one hand part of the absorbed energy of fabric having hexagonal mesh structure is consumed on the breakage of surface yarns, and the former be used to absorb more total energy than the latter. Two fabrics having pillar and weft insertion structure but with slightly varying thickness represent spacer fabrics having similar surface structure and needle bar distance, but different finenesses of spacer yarns. In the case of fabric having coarser spacer yarns the Fmax value of fabric is lower when compared to that with the finer ones. Hence the former can be used to absorb more energy than the latter. A high significance has been found regarding that the influences of surface structure, fabric thickness, and spacer yarns fineness of fabrics impacted under various energies on peak force and absorbed energy.

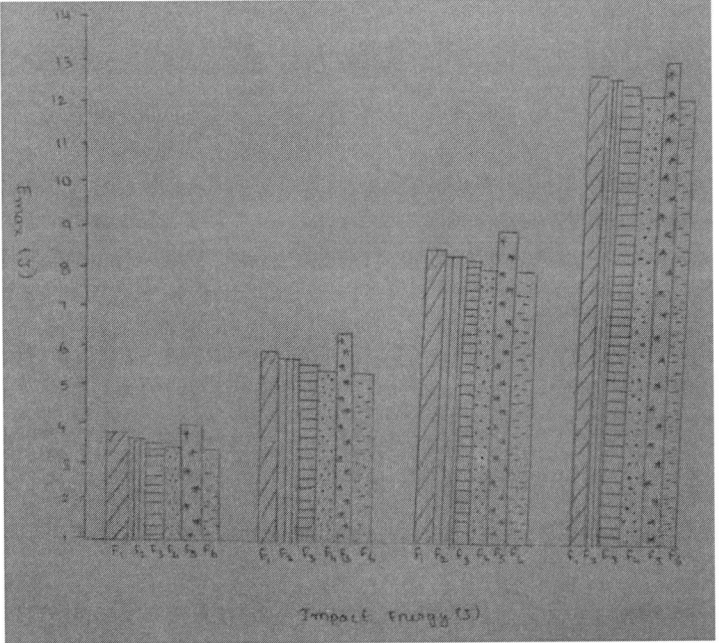

Figure 6 – Peak forces and absorbed energies for all tested fabrics[18]

8.5   Damage deformation and depth

Despite the same technique (flat-sphere) being used for all impacts, the damaged surfaces of test fabrics and their depth values differed from each other. But, deformation and depth show a fabric's ability to resist impact. Small damage depth illustrates good impact resistance in the same conditions. A special 3D scanner has been used to study the typical damaged surface deformations of specimens. It has been found that the surface deformations of fabrics having the same surface layers under four different impact energies resemble each other in shape but not in dimensions. In the case of fabrics having varied surface structures impacted under the same energy differ from each another. Also, there are many height lines surrounding the damaged areas for fabrics having rhombic and hexagonal mesh structures. Since the surface layers of fabric having pillar and weft insertion structure are plain and tight, whereas fabrics having rhombic and hexagonal mesh structures have many meshes on the outer layers, laser beams derived from the transmitter could irradiate the closed top surface of fabric having pillar and weft insertion structure, while parts of them irradiated directly to the inner or even the bottom of the fabrics having rhombic and hexagonal mesh structures due to the open structures on outer layers (Figure 7). The damaged deformation of fabric having pillar and weft insertion structure shows that the damaged zone tends to become localized and is more or less a round pit in shape. The deepest point occurs in the center of the pit [18]. The damaged area appearances of fabric having pillar and weft insertion structure and rhombic mesh structure appear similar (a round pit). But, in the case of fabric having hexagonal mesh structure, the hexagonal sides have been broken. Photographs reveal that the outer layers of fabrics having pillar and weft insertion structure and rhombic mesh structure are still fine. On the other hand the appearance of fabric having hexagonal mesh structure reveals the breakage of hexagonal sides and collapse of spacer yarns around them, as greater stress concentrates on the mesh sides and parts of spacer yarns are exposed in the open air. Damage depths for all the test fabrics can be calculated based on the laser scanner. The values of depth of various test fabrics impacted under the same energy and those of the same test fabric (having pillar and weft insertion structure) subjected to various impact energies have been measured. The depth of fabric having pillar and weft insertion structure is smaller than that of fabric having rhombic mesh structure, which implies that the impact resistance of fabric having a closed surface layer is better than that having small-size meshes on surface layers. On the one hand, the damage depth of fabric having hexagonal mesh strucuture(with large-size meshes on the surface layers) is not matched with that of fabric having pillar and weft insertion structure and rhombic mesh structure, since their damage conditions vary, the surface yarns in the case of fabric having hexagonal mesh structure are broken while those having pillar and weft insertion structure and rhombic structure are not. Almost all absorbed energy is converted into irrecoverable deformation energy of spacer yarns. Hence, a pit is created on the fabric surfaces in the case of fabrics having pillar and weft insertion structure and rhombic mesh structure. On the one hand, in the case of fabric having hexagonal mesh structure part of the absorbed energy is consumed on the breakage of surface yarns, whereas the depth of damage zone decreases. Considering this aspect, fabric having hexagonal meshes is not a perfect protector since the surface is easy to be damaged, and that large-size meshes on outer layers cannot protect spacer yarns in the middle layer effectively. The depths of fabrics having greater thickness and coarser spacer yarns are less than those of their opposites.

The findings show that the former exhibit better impact resistance. The values of depth of fabric having pillar and weft insertion structure subjected to various energies have been determined. Thus, with the increase in impact energies there is increase in the depths. Further, depth values of fabrics under two lower impact energies approximate to each other, and those under the two others are similar, while the latter is larger than the former. The findings clearly indicate that the damage of fabrics impacted under greater energies is more serious compared with those under the lower ones.

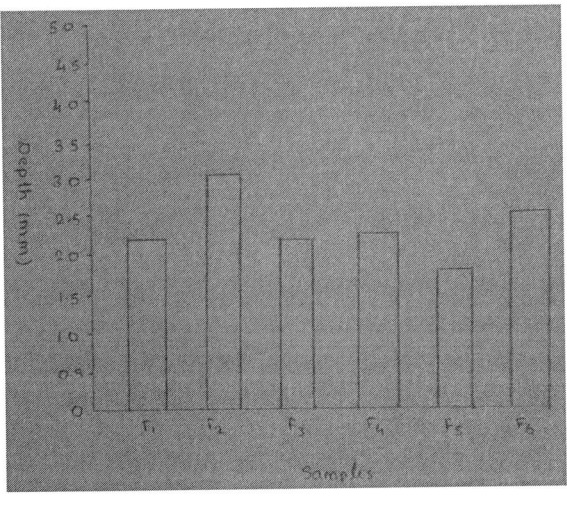

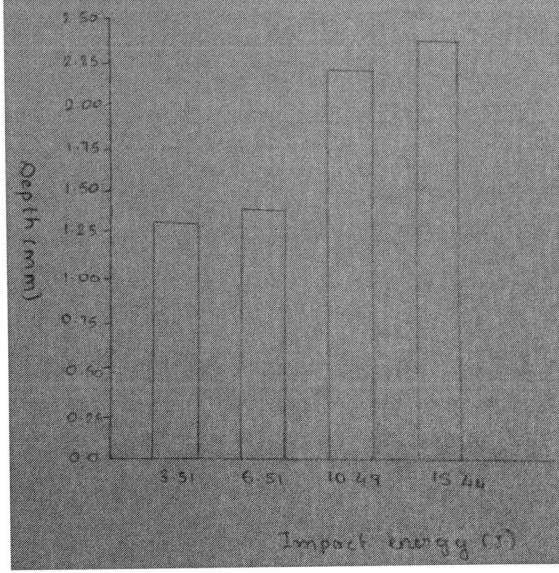

Figure 7 – Depths of damage of all test fabrics[18]

8.6   Compression-after-impact behavior

In the case of fabrics having pillar and weft insertion structures the typical pressure-compressibility curves before and after impacts are depicted in Figure 8. The curve depicted as 'OJ'represents that of the test fabric before impacts (undamaged fabric), while the others indicate those of fabrics after impacts under various energies [18]. It can be observed that the two types of curves, before and after impacts, have same configurations but varying magnitudes. All the curves can be split into two phases, namely, initial phase (first phase) and elastic phase (second phase), based on slope changes of the curves. Further, compression resistances of fabrics before and after impacts are not obviously different in first phase. On the other hand while in second phase, those of the damaged fabrics decline by a large margin. As the compression in first phase derives from fastening the multifilament loops and monofilaments, whereas that in second phase is mainly because of bending of spacer yarns, unfortunately, the spacer yarns have been damaged and their load-carrying capacities have diminished after impacts.

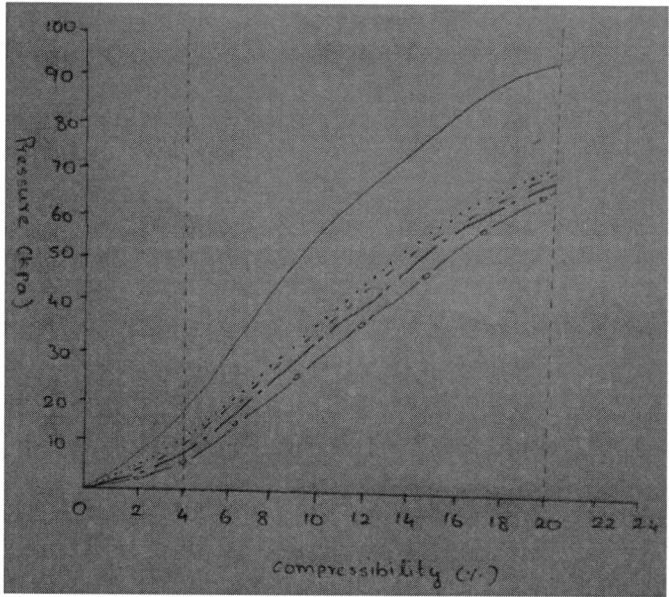

Figure 8 – Typical curves depicting pressure compressibility before and after impacts for F1 [18]

8.7   Residual strength

The residual strength is normalized in relation to damaged and corresponding undamaged compression strength of each type of test fabric under 20% compressibility. It is graphically shown as a function of impact energy in Figure 9.

The related test results have been determined. It has been found that that undamaged fabrics having smaller size meshes on outer layers, smaller fabric thickness, and coarser spacer yarns exhibit better compression resistance [18]. But, residual strength values for each test fabric under different impact energies (except 0J) differ within a small range. In other words, these values incline to be stable, and damage of a specimen impacted under relative smaller energies tends to saturation. Hence, the average residual strength value of each test fabric in different impact energies is used to evaluate the compression resistances after impacts. It has been found that fabric having better compression resistance before an impact performs better after it as well. The drop-off rates (DO) of compressive strength in the case of all the the test fabrics can be calculated.

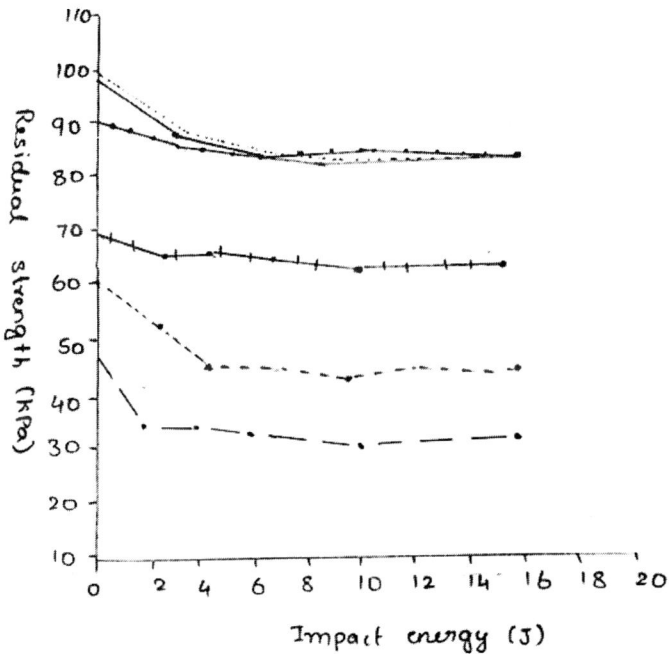

Figure 9 – Residual strength values of damaged and undamaged test fabrics against impact Energy [18]

It shows that fabrics having better compression resistance also exhibit greater drop-off rates. In the case of the fabrics having similar thickness, greater drop-off rates of fabrics (closed surface structure and coarser spacer yarns) imply that such fabrics can absorb more energy when impact occurs by forming irrecoverable deformation to reduce peak force. In the case of fabrics having varied thicknesses, those having greater thicknesses can absorb more energy and have lower drop-off rates. Considering the warp knit spacer fabrics as a cushion material, the impact behavior and damage characteristic after impacts have been studied. In order to study the behaviors of the spacer fabrics having various structures, 4 factors have been considered including peak force, absorbed energy, damage depth, and drop-off rate of residual strength. A spacer fabric having good protection performance was considered to have lower peak force, more absorbed energy, lower damage depth, and a lower drop-off rate of residual strength. Based on study of the experimental results, the conclusions arrived at have been summarized as below:

a) During an impact the variation of force arises due to the combined effect of two factors-displacement and failure of spacer yarns in the center zone. Also, force increases with the increasing of the former and reduction due to the latter.

b) In the case of fabrics having similar needle bar distance (various surface structures and finenesses of spacer yarns), fabrics having coarser spacer yarns and close surface structure show several satisfying results, like lower peak force, larger absorbed energy, and lower depth value, but higher drop-off rate of residual strength. When taking into account the particular impact technique (flat-sphere), fabric having a mesh structure is body-fitted, since meshes can open and shut more freely depending on the shape of wrapped objects. But, meshes which are too large on surface layers renders the mesh sides inclined to collapse and implies that spacer yarns cannot be protected perfectly.

c) In the case of fabrics having varied distances of needle bar (similar surface structure and fineness of spacer yarns), fabric having greater thickness performs perfectly in the experiments (lower peak force, more absorbed energy, lower damage depth, and lower drop-off rate), but too thick a protector may cause discomfort to a person's body. It is necessary to take into account the balance between the protective performance and comfort through choice of an appropriate fabric thickness for the particular protective end use.

d) In the case of a fabric subjected to various energies, as the energy levels increase the peak force, absorbed energy, and damage depth increase, whereas residual strengths approximate to each other.

8.8   References

[1]   Liu YP and Hu H. Compression property and air permeability of weft-knitted spacer fabrics. J Textil Inst 2011; 102: 366–372.

[2]   Liu YP, Hu H, Zhao L, et al. Compression behavior of warp-knitted spacer fabrics for cushioning applications. Textil Res J 2012; 82: 11–20.

[3]   3.Ye XH, Hu H and Feng XW. Development of the warp knitted space fabrics for cushion applications. J Ind Textil 2008; 37: 213–223.

[4]   Ye XH and Hu H. Application of warp-knitted space fabric in car seats. J Textil Inst 2007; 98: 337–344.

[5]   Bruer SM and Smith G. Three-dimensionally knit spacer fabric: a review of production techniques and applications. J Textil Apparel Technol Manag 2005; 4: 23–25.

[6]   Yip J and Ng SP. Study of three-dimensional space fabrics: physical and mechanical properties. J Mater Process Technol 2008; 206: 359–364.

[7]   Liu YP and Hu H. Sound absorption behavior of warp knitted spacer fabrics. Textil Res J 2010; 80: 1949–1957.

[8]   Chen Y, Jiang GM and Chen HX. Research on the pressure relief of warp knitted spacer fabric. Knitting Ind 2008; 4: 26–28.

[9]   Ye XH, Hu H and Feng XW. Research on pressure relief of warp knitted spacer fabric. J Donghua Univ 2006; 32: 88–91.

[10]   Armakan DM and Roye A. A study on the compression behavior of space fabrics designed for concrete applications. Fiber Polym 2009; 10: 116–123.

[11]   Sheikhzadeh M, Ghane M, Eslamian Z, et al. A modeling study on the lateral compressive behavior of spacer fabrics. J Textil Inst 2010; 101: 795–800.

[12]   Miao XH and Ge MQ. Vibration behavior of cushioning warp knitted spacer fabric. J Textil Res 2008; 29: 57–60.

[13]   Chen Y. Compression resistance of warp knitted spacer fabric. Dissertation, Jiangnan University, 2007, p.19.

[14]   Chen HL. Development and research on two-bed Raschel spacer fabrics. Dissertation, Donghua University, 1998.

[15]   Hou XN, Hu H, Liu YP, et al. Nonlinear compression behavior of warp-knitted spacer fabric: effect of sandwich structure. Textil Res J 2011; 23: 119–134.

[16]   Liu YP, Hu H, Long HR, et al. Impact compressive behavior of warp-knitted spacer fabrics for protective applications. Textil Res J 2012; 82: 773–788.

[17]   Douglas P, Magaziner J, Zimmerman S, et al. Efficacy of a hip protector to prevent hip fracture in nursing home residents. J Am Med Assoc 2007; 298: 413–422.

[18]   Xiaofang G, Hairu L and Li Z, Investigation on the impact and compression-after-impact properties of warp-knitted spacer fabrics, Textile Research Journal, 83(9) 904–916,

Chapter 9

Analysis of impact compressive behavior of warp knit spacer fabrics for evaluation of suitability in protective applications

Summary

A drop weight tester has been used to test a number of warp-knitted spacer fabrics having predefined impact energy. A low pass filter has been used to determine the transmitted force signals as well as the acceleration. Comparisons have been made between the peak contact force and transmitted force as a function of time. Also the velocity and displacement of the striker have been calculated. Considering the contact force-displacement curve and energy absorbed-contact force curve the impact compressive behavior of a typical fabric has been evaluated. The influences of various structural factors, such as the spacer monofilament yarn inclination and fineness, fabric thickness, and surface knitted structures, on the impact compressive behavior of the warp-knitted spacer fabrics have been evaluated considering the contact force-displacement curves, energy absorbed-contact force curves and transmitted-time curves. A relation has been established between the peak transmitted and contact force. The investigations have revealed a number of interesting results. In the case of a warp knit spacer fabric the impact process is of a highly complex nature. For better analysis of the impact process in the fabric the acceleration of the striker as well as the transmitted force as a function of time are considered to be crucial factors which require to be measured. There is an impacting phase and rebounding phase involved in the entire impact process. During their peak time and value there are differences in peak contact and transmitted force. Because of shock wave propagation the highest point is reached by peak transmitted force. The peak transmitted force is much lower than the peak contact force. The impact behavior of spacer fabrics could considerably be affected by the inclination of the spacer yarn. Greater peak contact and transmitted force can arise from too vertical and inclined spacer yarns. Because of lower energy absorption capacity at the end of the plateau stage there in increase in peak contact and transmitted force with reduction of the spacer monofilament fineness. Compared to other structural factors, the influence of the fabric thickness is more complex. At higher fabric thickness there is reduction in fluctuations of the contact force. A better protective performance having peak contact and transmitted force can be achieved by means of an optimized fabric thickness. On comparison with fabrics having small mesh size and close structures, the fabrics having large-size mesh structures for both top and bottom surface layers exhibit greater peak contact and transmitted force. The relation between the peak contact and transmitted force is linear. For all spacer fabrics investigated the transmitting rate is almost same and is about 66.84%, irrespective of their structural factors. Considering the high energy capacity and lower peak transmitted force, the knit spacer fabrics are suitable as effective material through selection of appropriate structural factors for human body protection, like shoulder and back protectors in sports or other severe occasions.

9.1   Introduction

During the recent years there has been a major focus on the use of warp-knitted spacer fabrics as cushioning material in personnel protective clothing and equipment against impact [1–3]. Besides the necessity of comfort, the energy absorption ability and impact force attenuation of the fabric is also considered crucial for protection of the human body from injuries [4]. Earlier investigations on the compression and energy absorption have enabled to establish that the warp-knitted spacer fabrics are ideal cushioning materials for these kinds of application [5]. But, only under static testing condition the results of these investigations have been obtained. Moreover, literature shows that many other experimental and theoretical investigations relating to compressive behavior of the spacer fabrics have concentrated only on the static testing conditions [6-13]. As regards the dynamic compressive behavior only a small number of investigations have been carried out on warp-knitted spacer fabrics under impact. In the case of static as well as dynamic testing conditions, it has been established that the compressive behavior of a cushioning material can vary [14]. The impact object is subjected to a period of deceleration during impact process and is determined by the predefined impact energy and dynamic compressive behavior of the impacted material [15,16]. The peak contact force will be generated on the impacted side of cushioning material, and this force will partially be transmitted to the other side of the material during the impact process. The peak contact as well as transmitted force has direct relation to the maximum acceleration of the striker [17]. In order to achieve effective protection of human body, the peak force transmitted to a specific body part in the event of impact should not go above the endurance limit of tissue or bones, which means that the cushioning material should dissipate the kinetic energy of the impacting mass and maintains the peak force transmitted below the limit of the specific body part. The contact and transmitted force are not the same under the impact state. It differs from the static compression, wherein there is zero acceleration. The forces of compression applied on either sides of the test material are identical during a static compression test. But, the contact force applied on the impacted side of the material by the striker differs from the transmitted force to the other side during a dynamic compression test. Hence, the impact compressive behavior of a protective material and its performance can be better understood by the analysis of the contact force as well as transmitted force and their relation. Warp-knitted spacer fabrics are formed with a sandwich structure in which two surface fabric layers are connected together by a layer of spacer monofilament yarns. Such structural feature renders them easily tailor able for satisfying special needs in various protective applications by altering their structural factors. Factors such as the surface layer knitted structure, fabric thickness and density, spacer monofilament fineness and inclination, surface yarn type, linear density, and so on can be altered in a warp-knitted spacer structure. Previous findings reveal that such structural factors can significantly influence the compressive properties of warp-knitted spacer fabrics. This chapter highlights an investigation of the impact compressive behavior of the warp-knitted spacer fabrics for protective applications. The fabrics have been tested with a predefined impacting energy using a drop tester. Depending on the impact contact force-displacement curve, energy absorbed-contact force curve, and transmitted force-time curve the impact process of a typical spacer fabric has been evaluated. The influences of the structural factors like spacer yarn inclination and fineness, fabric thickness, and surface layer structure, on the impact compressive behavior and protection performance of the fabrics have also been explained. The investigation can possibly enable a better understanding of the impact behavior of warp-knitted spacer fabrics and could provide optimization of their structural design for human body protection.

9.2   Technical details

In order to analyze the influences of structural factors on the impact compressive of spacer fabrics, warp-knitted spacer fabrics have been knitted having varied fabric thicknesses, surface structures, spacer monofilament fineness, and inclinations [18]. A double needle bar raschel warp knitting machine has been used for knitting the fabrics. The surface layers have been knitted adopting 4 different structures, which include locknit, chain plus inlay, small-size rhombic mesh, and large-size hexagonal mesh. The two surface layers have been connected by 3 different spacer yarn inclinations that respectively correspond to shifting one, two, and three needle distances between the front needle and back needle bars (Figure 1).The surface layers the fabrics have been knitted using polyester multifilament, the spacer layer has been knitted with polyester monofilament of two different diameters for different fabrics.

(a)   (b)   (c)

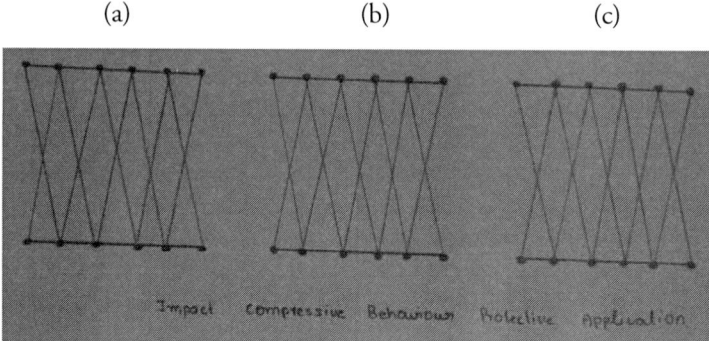

Figure 1 – Inclination of spacer yarn[18]
a)   Shifting a distance of single needle
b)   Shifting a distance of double needle
c)   Shifting a distance of triple needle

The dynamic compression tests have been carried out on a drop-weight impact tester working on the principle of low energy at low speed. It is able to measure the changes in acceleration of the drop striker and the force transmitted from the top side to the bottom side of the test fabric. A load cell has been used to measure the transmitted force. An ISOTRON accelerometer has been used to measure the acceleration of the drop striker. The impact signals have been detected and amplified by two identical charge amplifiers. The transmitted force and acceleration have been recorded by a high-speed data capture card. A strain gauge conditioner has been used to control the distance from the drop striker to the anvil. Fabrics have been cut in circular form. The test fabrics have been cut 7 times by using laser beam. In order to avoid the possible movement of the specimen during testing, its bottom surface was secured to the anvil.

One way analysis of variance has been used to determine the peak contact forces and peak transmitted forces for each dependent variable.

9.3   Study of impact process

The impact process of the spacer fabric has been studied adopting the following steps [18]

a)   Treatment of the acceleration and transmitted force signals

b)   Comparison of the contact force and transmitted force

c)   Calculation of velocity and displacement of the striker

d)   Impact compressive behavior analysis

e)   Effects of structural parameters

9.4   Evaluation of impact process

Treatment of the acceleration and transmitted force signals

Considering the test fabric (formed by locknit structure for top and bottom layers with spacer yarn inclination caused by shifting two needle distance) as an example, its acceleration and transmitted force signals as a function of time from one test in the original form as well as the form after filtering using a low-pass filter through Fast Fourier Transform at a set cutoff frequency of 5000 Hz have been depicted in Figure 2. The cutoff frequency has been determined based on a specific technique, which showed that a cutoff frequency of a half-sine waveform should be at least five times of the fundamental frequency of the shock pulse. The technique also showed that the pulse duration should be measured between the point at 10% of the peak acceleration during the rising time and the point at 10% of the peak acceleration during the duration of decaying. As the obtained signal was not exactly half-sine waveform, only the peak wave has been taken into account and its time is nearly 1.2 ms. Under such condition, the cutoff frequency should be above 4166 Hz. For satisfying this need and to achieve a good presentation, 5000 Hz has been used for the entire investigation. The low-pass filter with this cutoff frequency is found to be effective to eliminate the noise with a high accuracy.

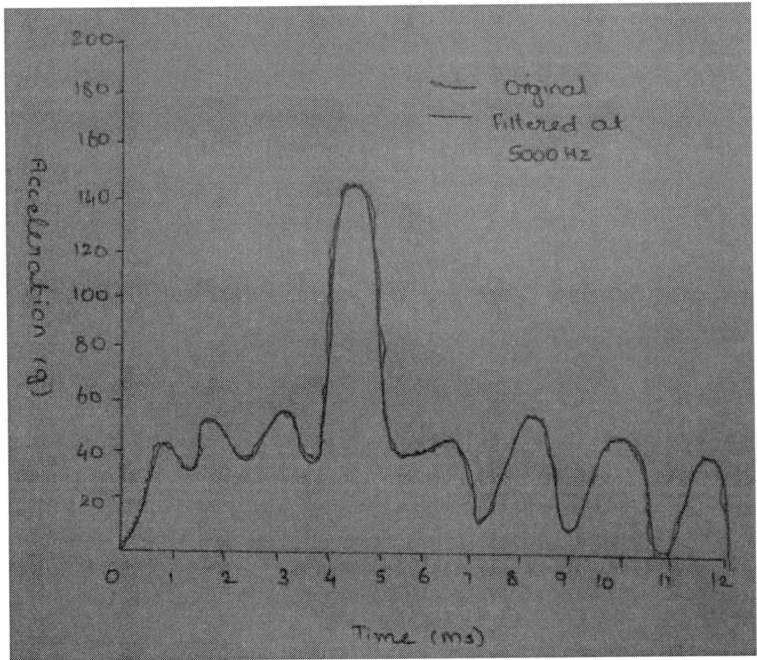

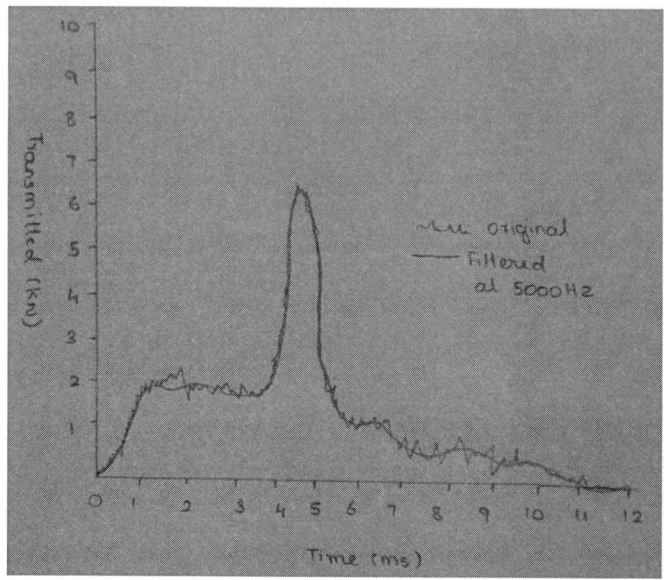

Figure 2 – Signals in original and after filtering for test fabric(formed by locknit structure for top and bottom layers with spacer yarn inclination caused by shifting two needle distance)[18]

a)   Acceleration
b)   Transmitted force

9.5   Contact force vs transmitted force

The contact force is proportional to the acceleration and is calculated based on the Newton's second law. For enabling better understanding the impact process, the contact force as well as transmitted force in the case of test fabric have been plotted together against the time (Figure 3). The contact force attains its peak point after about 4.3 ms of the impact and the impact process completes within about 12 ms. Depending on the theorem of momentum, a longer duration of the impact leads to a lower acceleration and a lower dynamic force. In the absence of the fabric, the motion of the drop striker is instantaneously stopped by the anvil, and hence results in a very high acceleration and dynamic contact. But, when the fabric is fixed on the anvil, the drop striker gets decelerated as it strikes the specimen. During this process of deceleration, the fabric receives and stores the kinetic energy of the impact and then releases it over a longer duration, thus resulting in a decrease in acceleration of the striker and generating a smaller contact force between the striker face and the top layer of the fabric. Also, inherent damping of the fabric can dissipate the kinetic energy and then result in further decrease in the contact force or acceleration. The contact force is transmitted as shock wave propagation.

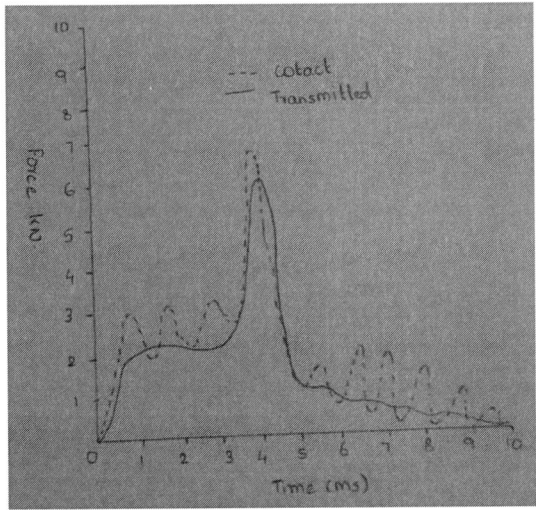

Figure 3 – Contact force versus transmitted force for test fabric(formed by locknit structure for top and bottom layers with spacer yarn inclination caused by shifting two needle distance)[18]

The highly complicated impacting process might include buckling, shearing, rotating, inter-contacting, and collapsing of the spacer monofilament yarns, as well as their contacts with the surface layers. These deformations contribute to the overall elastic-plastic deformation of the warp-knitted spacer fabric, which can store and dissipate the kinetic energy of the impact at the same time. It can be observed that the contact force and the transmitted force do not reach their peak points simultaneously. The transmitted force reaches its peak point with a time delay of 0.2 ms due to shock wave propagation. In addition, the peak transmitted force is much lower than the peak contact force. It should be pointed out that the shock wave propagation through the discrete fabric is very complicated and the energy loss during the wave propagation contributes to the lower transmitted force.

From Figure 5, it can be also found that the fluctuations exist in the contact force curve. It is believed that these fluctuations originate from the shearing and rotating deformations of the spacer monofilament yarns. It can be seen that these fluctuations are mitigated after transmission to the anvil.

9.6   Calculation of velocity and displacement of the striker

The velocity and displacement of the striker can be calculated by integrating the acceleration-time curve based on equations. The curves of the velocity and the displacement for warp knit spacer fabric(having locknit structure for top and bottom layers with spacer yarn inclination caused by shifting two needle distance) as a function of time are shown in Figure 4. It can be seen that the impact process can be split into two phases: impacting and rebounding. The impacting phase corresponds to a deceleration process from the start point when the striker starts to contact the fabric at the highest impact velocity (1.98 m/s) to the rebounding point when the velocity reaches 0 m/s. At the rebounding point, both the acceleration and the displacement reach their maximum values. While the impacting phase is related to the compression process of the fabric, the rebounding phase is related to the recovery process of the fabric. It is evident that the rebounding intensity largely depends on the recovery capacity of the spacer monofilament yarns, which function like springs in the compressed fabric.

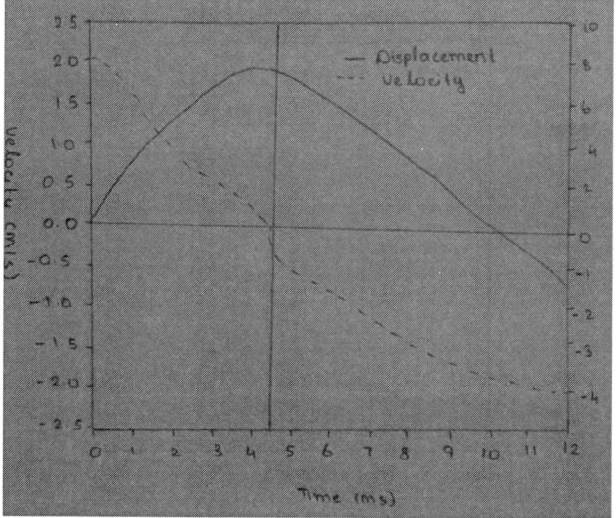

Figure 4 – Velocity and displacement as a function of time for test fabric(formed by locknit Structure for top and bottom layers with spacer yarn inclination caused by shifting two needle distance)[18]

9.7   Evaluation of Impact compressive behavior

In order to analyze the impact compressive behavior of the fabric, the contact force is plotted against the displacement, as shown in Figure 5. For the comparison, the compressive force-displacement curve of the same fabric obtained under the static compression condition (the compression speed is 12mm/min, and the fabric is compressed to 80% of its initial thickness) is also shown in Figure 5. It is found that the general trends of the two curves are very similar. However, some differences can be found. On the static curve, a very obvious plateau stage can be observed. However, on the impact curve, the fluctuations dominate in this stage. This phenomenon has not yet been clearly explained. The reason may come from the shearing and rotating deformations of the spacer monofilament yarns, as explained before. In addition, the impact contact force is higher than the static force during almost the entire compression process. For the impact test, the fabric was compressed to the largest displacement of 5.74mm with a peak contact force 7.83 kN. The static force at this displacement with a value of 5.25 kN is much lower than the contact force under impact. It is evident that the spacer fabric inherently exhibits strain rate sensitivity, which is similar to the polyester filaments used as spacer yarns. The monofilament spacer yarn inertia to rotation and deformation especially at high strain rates could also play a significant role. The rate dependency of the spacer fabric leads to a higher energy absorption capacity in impact than that in static compression. The energy absorbed-displacement curves are also shown in Figure 5. It is confirmed that the energy absorbed in impact is higher than that in static compression. At the point of peak contact force or the largest displacement in impact, the energy absorbed is 12.74 J. At the same displacement under static compression, the energy absorbed is much lower, with a value of 9.2 J.

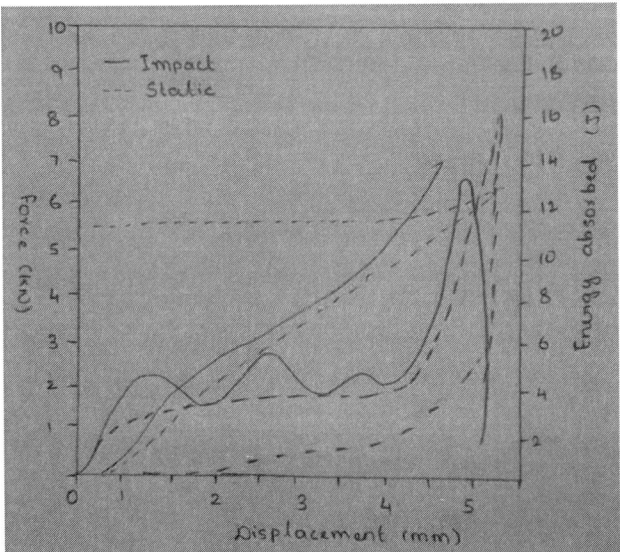

Figure 5 – Force and energy absorbed versus displacement curves under impact and static compression for test fabric (formed by locknit structure for top and bottom layers with spacer yarn inclination caused by shifting two needle distance)[18] Thick lines represent force and thin lines represent energy absorbed.

The energy absorbed within the plateau stage and the relevant plateau force are two key parameters to select and optimize the fabric for impact protection. To better understand the protective performance of the spacer fabric under impact, the energy absorbed by the spacer fabric is plotted against the contact force, as shown in Figure 6. It is found that the spacer fabric can absorb the majority of the impact energy before reaching the peak contact force. This can be confirmed by comparing the energy absorbed at point A, from which the contact force starts rising to its peak point B, to the total energy absorbed. At point A, the energy absorbed is about 10.5 J, which represents about 82.5% of the total energy absorbed (12.74 J). For comparison, the energy absorbed-force curve under static compression is also included in Figure 8. It is observed that the force under static compression is lower than the contact force under impact for the same level of energy absorbed from the beginning to the point at about 7.5 J. After this point, the static force becomes much larger than the contact force. At the end of the static compression with a displacement 6.02 mm, the energy absorbed is 11.18 J, which is lower than the energy absorbed in the impact test (12.74 J) at its maximum displacement of 5.74 mm. This result shows that the warp-knitted spacer fabric is an effective material for absorbing impact energy for protective application. Moreover, the energy-absorbing performance in terms of energy-absorbing capacity and contact force is rate dependency.

The above analysis shows that the compressive behavior of the warp-knitted spacer fabric under impact is different from the behavior under the static situation. The following discussions will focus on the effects of fabric structural parameters on the impact compressive behavior of the warp-knitted fabrics based on their impact contact force-displacement curves, energy absorbed-contact force curves, and transmitted force-time curves.

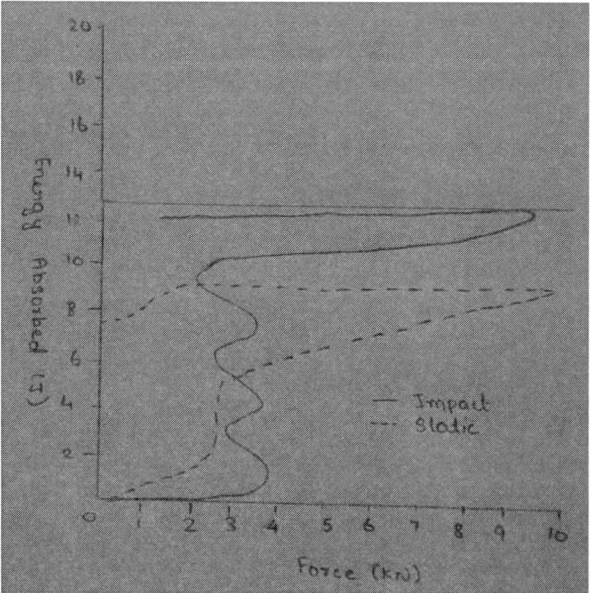

Figure 6 – Energy absorbed versus force curves under impact and static compression for test fabric (formed by locknit structure for top and bottom layers with spacer yarn inclination caused by shifting two needle distance)[18]

9.8   Influences of fabric structural factors

Influence of the spacer yarn inclination

The influence of the spacer yarn inclination on the impact compressive behavior of warp knit spacer fabrics has been studied using a group of three fabrics having the same surface layer structure (chain plus inlay) and the same spacer monofilament yarn but with different spacer yarn inclination (shifting one needle, two needles, and three needles distance between the front and back needles) [18]. The fabric thickness and stitch density of the surface layers have been maintained almost equal. As found from earlier study the spacer yarn inclination as well as length depends on the needle distance shifted between the front and back needles, and they increase with increasing the number of the needles shifted. Figures 7(a)–(c) depict the contact force-displacement curves, energy absorbed-contact force curves, and transmitted force-time curves of these fabrics.

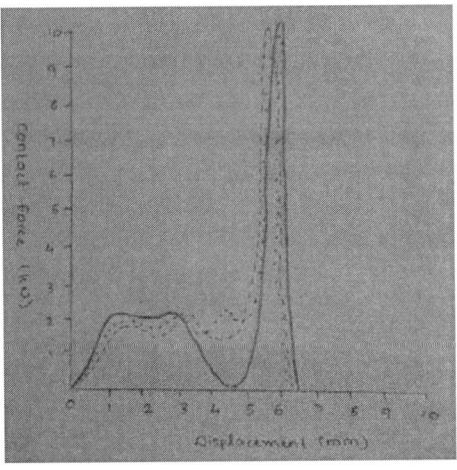

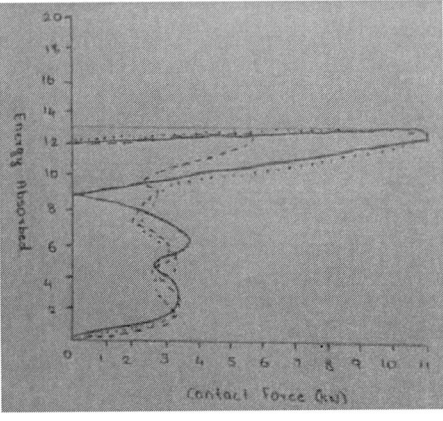

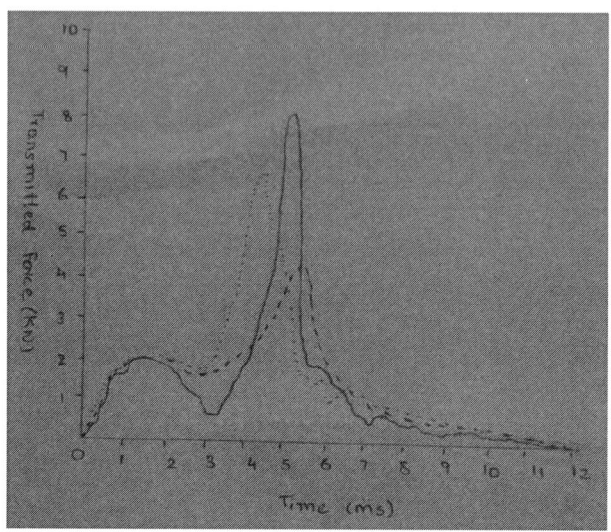

Figure 7 – Influence of spacer yarn inclination on impact compressive behavior[18]

    a)   Contact force-displacement curves
    b)   Energy absorbed –contact force curves
    c)   Transmitted force –time curves

Their peak contact force and peak transmitted force have been determined. These figures reveal the following phenomena

    a)   As the spacer yarn inclination reduces the average level of the contact force at the plateau stage up to 3.5mm of displacement increases as shown in Figure 7(a). But, at the end of the plateu stage there is a clear fall in the contact force for the spacer fabric with only shifting one needle distance of spacer yarns.

    b)   As shown in figure 7(c), similar trend has also been observed from the transmitted force-time curves. In this case, as the spacer yarn inclination reduces within 2 ms of the impact, the average level of the transmitted force increases. By shifting the spacer yarns a distance of a needle there is also a marked fall in the transmitted force for the spacer fabric.

    c)   By shifting the spacer yarns a distance of 1 and 3 needles the peak contact forces and peak transmitted forces of the fabrics the peak contact as well as transmitted forces are far greater compared with those of the fabric where the spacer yarns are shifted a distance of two needles. The findings of one-way ANOVA show good significance relating to the influences of spacer yarn inclination in the case of both peak contact force as well as peak transmitted force. But, by shifting the spacer yarns a distance of one and three needles the peak contact forces of the fabrics remain nearly the same. In the case where spacer yarn is shifted a distance of one needle the peak transmitted force of the fabric is greater than that of shifting a distance of three needles. By studying the ability of the spacer yarns to endure the contact force during the impact process it is possible to explain these phenomena. Since the spacer yarns having lower inclination of the spacer yarns are more oriented to the impact direction, as the inclination of the spacer yarns reduces the compression resistance of spacer fabrics increases.

Based on the surface structure tightness the boundary conditions of the spacer yarns could be varied from the pinned constraint to the fixed constraint as the ends of the spacer monofilament yarns are bound by the multifilament stitches in the surface layers. The spacer yarn having lower inclination has the shorter length as the thicknesses of the fabrics of the test group are almost the same. Hence in the event of impact compression the fabric having lower inclination will have higher compression resistance. But, the structure of the fabric becomes less stable when the spacer yarns are too vertical to the surface layers (shifting one needle distance), resulting in shearing between the two surface layers, since under impact loads the spacer yarns tend to tilt along the course direction. Observation of the cross sections of the test fabrics confirms the phenomenon. It has been found that while the channels are formed by spacer yarns in the cross-section of test fabric formed by shifting one needle distance of spacer yarns, the intersection of the spacer yarns are found in test fabrics formed by shifting two and three needle distances of spacer yarns, respectively. The channels formed in test fabric with shifting two needle distance result in the periodic distribution of the compression stress onto the surface layers under impact. Upon application of uneven stress distribution the shearing between the two surface layers along the course direction can easily occur. This attributes an obvious drop of the contact force at the end of plateau stage is found in the case of test fabric having one needle shifting of spacer yarns. Whereas, the spacer yarn can get easily crushed under the contact force if they are too inclined. As a result, the fabric will absorb less energy at the plateau stage. Simultaneously, under impact the fabric having more inclined spacer yarns can be more easily densified. Greater peak contact force arises due to lower energy absorption capacity at the plateau stage and higher densification of the fabric having more inclined spacer yarns. But, it is not easy for the shock wave to penetrate through the thickness of the fabric since there are more spacer yarns within the fabric knitted by shifting distance of three needles distance. In this connection, its transmitted peak force is lower than that of the fabric knitted with shifting one needle distance. The discussion clearly reveals that the spacer yarn inclination can significantly affect the impact behavior of spacer fabrics. Owing to greater peak contact and transmitted forces both the fabrics having lower inclination of spacer yarns (shifting one needle distance) and higher inclination of spacer yarns (shifting three needles distance) are unsuitable for protective applications because of higher peak contact and transmitted forces.

9.9   Influence of fineness of the spacer monofilament

The influence of fineness of spacer monofilament on the impact compressive behavior of the spacer fabrics have been investigated by selection of two fabrics having similar spacer yarn inclination (shifting three needles distance) and surface layer structure (chain plus inlay) but with two different spacer monofilament diameters [18]. Almost the same thickness and surface layer stitch densities of these fabrics have been maintained. The contact force-displacement, energy absorbed-contact force, and transmitted force-time curves of the spacer fabrics, as well as their peak contact and transmitted force have been measured. The following inferences have been can be made from the studies

a) With the increase in the fineness of the spacer monofilament the average level of the contact force at the elastic and plateau stages increases. It can be attributed to the various moments of inertia of the spacer monofilament with different radii. The cross-section of monofilaments is circular. It has been found the radius has a considerable influence on the buckling load under compression.

b) Owing to the greater force level at the elastic and plateau stages the fabric having coarser spacer monofilament can absorb more energy at these stages compared to the fabric having finer spacer monofilament. The fabric having finer spacer monofilament is capable of absorbing more energy at the densification stage. Hence, during the densification stage the contact force of the fabric having the finer spacer monofilament is far greater compared to the coarser spacer mono filament. The increase of the peak contact and transmitted forces with decreasing fineness of spacer monofilament is attributed to various level of densification. The influence of fineness of spacer yarn on the peak contact and transmitted force are highly significant as showd by ANOVA test results.

c) In the case of the fabrics having different spacer monofilament fineness, the contact and transmitted forces of the fabrics do not attain their peak points simultaneously. Under impact the initial peak contact force point and peak transmitted force point are attained by the spacer fabric having the finer spacer monofilament yarns. It is due to the higher acceleration at the densification stage of the fabric having finer spacer monofilament yarns which can stop the striker in a shorter duration. The period of deceleration is necessary to the impact protection, as the impact process should be as long as possible to minimize peak acceleration, which is directly related to the peak contact force and peak transmitted force.

The aforesaid discussion shows that the decrease of spacer monofilament fineness can considerably reduce the energy absorption capacity of spacer fabric and protective performance. But, with the increase in the spacer monofilament fineness the fabric stiffness increases, and hence reduce the comfort property of the fabric. Hence, it is necessary to consider the balance between the comfort and protective performance through choice of appropriate spacer monofilament fineness for a particular protective end use.

9.10   Influence of fabric thickness

In order to study the influence of fabric thickness on the impact compressive behavior of warp knit spacer fabrics a group of four fabrics have been knitted having similar spacer monofilament diameter, spacer yarn inclination (shifting two needles distance), and surface layer structure (chain plus inlay) but having various thicknesses [18]. The contact force-displacement curves, energy absorbed-contact force curves, and transmitted force-time curves of these fabrics have been measured. Their peak contact force and peak transmitted force have also been measured. Studies have revealed the following phenomena.

a)   As the fabric thickness increases there is reduction in the fluctuations of the contact force-displacement curves before rising to the peak force. In the case of thicker fabrics the obvious plateau stage is achieved.

b)   As the fabric thickness increases there is reduction in the average level of contact force at the elastic and plateau stages. However, as the thickness increases the displacement at the plateau stage increases.

c)   As compared to the thicker fabrics the thinner fabrics can absorb more energy before starting to rise to the peak force. However, this occurs at a higher contact force level.

d)   Similar trend has been observed from the transmitted force-time curves. In this case, during the elastic and plateu stages, as the fabric thickness increases, there is reduction in the average level of the transmitted force.

e)   Initially there is reduction in the peak contact and peak transmitted forces followed by increase with increasing fabric thickness. A high significance is noticed with regard to the influence of fabric thickness on the peak contact force and peak transmitted force as indicated by ANOVA test results.

The reasons given below could used to explain all the above phenomena

a)   The spacer yarns become longer as the fabric thickness increases. When the spacer yarns are considered as slender rods, there will be reduction in their compression resistance with increase in their length. Hence, in comparison with thinner fabrics, the thicker fabrics will become softer and their compression force level at the plateau stage will become lower. Also under impact the softer fabrics get densified more easily and could result in a slow increase of the peak contact force at the densification stage as the fabric thickness increases. Whereas, in the case of a thinner fabric the decrease of the spacer yarn length renders it stiffer. The fabric can become very stiff upon decreasing the fabric thickness to a certain level. Under such circumstance, the reaction force between the fabric and the drop striker can considerably increase, and can lead to an increase of the peak contact force. Hence, in order to achieve a lower peak contact force there is an optimized fabric thickness.

b) In comparison with a thinner spacer fabric a thicker spacer fabric has more space between surface layers, and hence it can be expected that the displacement increases with increasing the fabric thickness. In the case of a thicker spacer fabric slight fluctuations at the plateu stage can be seen since the compression process of the fabric under impact becomes smoother due to the increase of fabric thickness. On the other hand, with a thinner fabric more vibrations can be created during the impact process that can lead to severe fluctuations in its contact force-displacement curve.

c) During the plateau stage the displacement as well as the contact force level determines the energy absorbed from the starting impact point to the end point of the plateau. For the thicker fabrics, although their displacements are higher, the lower contact force level reduces the amount of the energy they absorb from the starting impact point to the end point of the plateau.

But, the displacement can be can be reduced a thickness that is too small. It can lead to decrease of the total energy absorbed at the end of the plateau. Hence, in order to achieve greater energy absorption with lower force there is an optimized fabric thickness. The aforesaid discussions indicate that the impact compressive behavior of the spacer fabric is significantly affected by the fabric thickness. Through choice of appropriate fabric thickness by maintaining other structural factors constant it is possible to optimize the protective performance of a spacer fabric.

9.11   Influence of the layer of surface structure

The surface layer knitted structure affects the distribution, binding condition and inclination of the spacer monofilament yarns which are bound by the multifilament stitches in the surface layers [18]. Hence, in order to study the influence of the surface layer structure on the impact compressive behavior of the spacer fabric a group of six fabrics having similar needle distance shifted for the spacer yarns (two needles) and almost similar thickness but with different surface layer structures have been selected. The surface layer structures comprise of 3 subgroups as follows

a) Both the surface layers having a close structure

b) The upper layer having an open structure and the lower layer having a close structure, and

c) Both the surface layers having an open structure.

The close structures include locknit and chain plus inlay, and the open structures include small-size rhombic mesh and large-size hexagonal mesh. The contact force-displacement curves, energy absorbed-contact force curves, and transmitted force-time curves of the fabrics have been measured. Their peak contact force and peak transmitted force have been measured. The surface structures evidently influence the peak contact force, the energy absorbed, and the peak transmitted force. On the one hand the fabric knitted having large-size hexagonal mesh for both surface layers exhibit the greatest peak contact force and the peak transmitted force, and the lowest energy absorption capacity at the end of plateau stage. Whereas the fabric having small-size rhombic mesh for both surface layers exhibit the lowest peak contact force and the lowest peak transmitted force, and the highest energy absorption capacity at the end of plateau stage. In addition, when compared with the fabric having small size rhombic mesh for the top surface layer, the fabric having large-size hexagonal mesh for the upper surface layer has a higher peak contact force and peak transmitted force compared to those of the fabric having small-size rhombic mesh for the upper surface layer, although their bottom surface layers are knitted with the same chain plus inlay structure. In addition, the fabrics having the close structure for the upper as well as lower surface layers exhibit moderate peak contact force and peak transmitted force. With the close structures, the fabric having chain plus inlay exhibits a lower peak contact force and a lower peak transmitted force compared to those of fabric having locknit. Such phenomena can be understood through study of the distribution, binding condition, and inclination of spacer monofilament yarns. It is well known that the buckling load of an elastic rod is determined by its boundary condition and initial geometrical shape. Upon studying the microscopic photographs of surface layers, it can be observed that the fabrics having close structures show larger and looser stitches in the surface layers compared with those of the fabrics having open structures. Also, in comparison with fabric having locknit structure the stitches in the surface layers of the fabric having chain plus inlay are smaller. It indicates that the ends of the monofilament yarns are more tightly bound by the multifilament stitches in the fabrics having open surface structures compared with those having close surface structures. With the close structures, the monofilament yarns are more tightly bound by chain plus inlay than locknit. The initial shape of the spacer monofilament yarns is another parameter that determines the buckling load.

When the cross-sectional microscopic photographs of the fabrics are studied, it is observed that the large-size hexagonal mesh surface layer renders the spacer monofilament yarns highly inclined and buckled and results in the reduction of the compression resistance. Whereas, the initial shapes of spacer monofilament yarns among the fabrics having close and small-size mesh surface structures show no significant differences. When the boundary conditions and initial geometrical shapes of spacer monofilament yarns are taken into account, it is easy to understand that the fabrics having large-size mesh structures show the poorest impact protection performance because of the highly buckled and inclined spacer monofilament yarns, and the fabrics having small-size mesh show the best impact protection ability because of the tight binding structures. Owing to the combined effects of loose binding structures and lowly buckled and inclined spacer monofilament yarns, the fabrics knitted with a close structure have moderate impact protection performance. Also, in the case of human body protection the formability of the spacer fabrics is to be taken into account. The close structures render the fabrics stiffer and more difficult to fit the body shape with greater fabric thickness. But, the fabric having small-size mesh surface structure has better shapability and protective performance. It is possible to adjust the shape and size of the mesh structure on the surface layers of spacer fabric so as to satisfy the various needs of protection. Through choice of appropriate surface structure the balance between the shapability and protective performance needs to be considered for a particular protective application.

9.12   Relationship between the peak contact force and transmitted force

The crucial factor that decides the protective performance of a spacer fabric is the peak transmitted force. As discussed earlier, the peak force transmitted to a specific area of the body during the impact process should be above the endurance limit of tissue or bones. Hence, it is necessary to understand the extent to which peak contact force can be transmitted from the impacted side to the other side of a fabric. The peak transmitted force vs the peak contact force for all the test spacer fabrics have been studied. The peak transmitted force is known to have a very obviously linear relationship with the peak contact force. According to the linear regression calculation relating to the peak transmitted force and peak contact force about 66.84% of the peak contact force is transmitted from the impacted side to the other side of the fabric and this transmitting rate is almost equal in the case of all the warp-knitted spacer fabrics irrespective of their structural factors. The transmitted force decreases with the peak contact force. Because of this the optimized design of a spacer fabric should be focused on reduction of the peak contact force through choice of appropriate structural factors for a predefined impact energy.

9.13   References

[1]   Rock M and Lohmueller K. Three-dimensional knit spacer fabric for footwear and backpacks. Patent 5896758, USA, 1999.

[2]   Goodwin EL. Protective device using a spacer fabric. Patent 2006/0287622 A1, USA, 2006.

[3]   Sorensen BB. A device for protection of the hips. Patent EP1684604, Europe, 2007.

[4]   Chen XG, Sun Y and Gong XZ. Design, manufacture, and experimental analysis of 3D honeycomb textile composites, part II: experimental analysis. Textile Res J 2008; 78: 1011–1021.

[5]   Liu YP, Hu H, Zhao L, et al. Compression behavior of warp-knitted spacer fabrics for cushioning applications. Textile Res J. Epub ahead of print 2 August 2011. DOI: 10.1177/0040517511416283.

[6]   Ye X, Fangueiro R, Hu H, et al. Application of warp knitted spacer fabrics in car seats. J Text Inst 2007; 98: 337–344.

[7]   Ye X, Hu H and Feng X. Development of the warp knitted spacer fabrics for cushion applications. J Ind Text 2008; 37: 213–223.

[8]   Liu YP and Hu H. Compression property and air permeability of weft knitted spacer fabrics. J Text Inst 2011; 102: 366–372.

[9]   Miao XH and Ge MQ. The compression behavior of warp knitted spacer fabric. Fibres Text East Eur 2008; 16: 90–92.

[10]   Armakan DM and Roye A. A study on the compression behavior of spacer fabrics designed for concrete applications. Fiber Polym 2009; 10: 116–123.

[11]   Sheikhzadeh M, Ghane M, Eslamian Z, et al. A modeling study on the lateral compressive behavior of spacer fabrics. J Text Inst 2010; 101: 795–800.

[12]   Vassiliadis S, Kallivretaki A, Psilla N, et al. Numerical modeling of the compressional behavior of warp-knitted spacer fabrics. Fibres Text East Eur 2009; 76: 56–61.

[13]   Bagherzadeh R, Montazer M, Latifi M, et al. Evaluation of comfort properties of polyester knitted spacer fabrics finished with water repellent and antimicrobial agents. Fiber Polym 2007; 8: 386–392.

[14]   Miltz J and Gruenbaum G. Evaluation of cushioning properties of plastic foams from compressive measurements. Polym Eng Sci 1981; 21: 1010–1014.

[15]   Gruenbaum G and Miltz J. Static versus dynamic evaluation of cushioning properties of plastic foams. J Appl Polym Sci 1983; 28: 135–143.

[16]   Ramon O, Mizrahi S and Miltz J. Merits and limitations of the drop and shock tests in evaluating the dynamic properties of plastic foams. Polym Eng Sci 1994; 34: 1406–1410.

[17]   Lyn G and Mills NJ. Design of foam crash mats for head impact protection. Sport Eng 2001; 4: 153–163.

[18]   Yanping L, Hong H, Hairu L and Li Z, Impact compressive behavior of warp-knitted spacer fabrics for protective applications, Textile Research Journal 82(8) 773.

Chapter 10

Porosity and capillarity of weft knitted spacer structures

Summary

A number of fabric parameters affect the porosity of weft knitted spacer fabrics. The porosity can be calculated using the porosity model suggested. As the number of spacer yarns in a repeat, courses and wales per unit area increases, the porosity of weft knitted spacer structures increases. But, the change of tightness can be accommodated by the fabric thickness, thereby changing the porosity. In order to analyze the absorbency characteristics of spacer fabrics, the capillary radius model has been evolved. A formula is used to find the average capillary radius. Increase in the tightness of weft knit spacer structure can decrease the porosity. It leads to an increase in the number of courses and wales, and an increase in number of spacer yarns within a repeat. So too, change in the number of spacer yarns and tightness of the structure can change the radius of capillaries. Normally, with the decrease of capillary radius the number of spacer yarns in a repeat increase. However, since increase in spacer yarns leads to increase in fabric thickness, there is restriction in the decrease of capillary radius.

10.1   Introduction

Weft knitted tuck spacer structures comprise of upper and lower plain knitted layers which are joined by a mesh of yarn. Hence the structure of spacer fabric looks as a tight mesh or bundle of yarn sandwiched between two layers of plain knit fabrics. The upper and lower spacer fabric layers are by the mesh of yarn which is oriented at an angle between them. When compared with other textile structures like woven, braided and weft knits, the weft knit spacer fabrics are structurally more open in character. The porosity and the size of capillary of a knitted spacer structure will influence its physical properties such as the bulk density, liquid uptake, the mass transfer and the thermal conductivity. Mathematical models have been developed for prediction of the porosity and the capillary radius of knitted spacer structures. They would help to design knitted structures for achieving desired level of porosity and radius of capillary, and thereby determine other properties like absorbency, absorbency rate and liquid retention.

The 2D and 3D geometric models of knit structures have been studied and subsequently many researchers have studied knitted structures [1,2]. Both the models have shown the observed relationships between fabric parameters like the courses and wales per unit length, the stitch length and the yarn diameter. The dimensional properties like stitch length, fabric weight, fabric thickness and the courses to wales ratio have also been investigated using the 2D geometrical model and concluded that the theoretical fabric thickness, twice the yarn diameter, is close to the practical value that has been obtained [3]. The bulk density and the thickness of plain knitted fabrics has been studied and it has been suggested that the fabric thickness of a plain knitted fabric in its fully relaxed state is higher than that of its dry relaxed state [4]. Also, the relationship between geometrical thickness, fabric loop length and bulk density of plain knitted structures has been studied. A proportional relationship between the parameter thickness/ length and the fabric tightness has been studied, and this was interpreted in terms of the basic plain knitted fabric geometry. Also it has been noticed that there exists a linear relationship between the fabric bulk density, and the fabric tightness factor and the fibre specific gravity. A theoretical model has been proposed for prediction of the porosity of a plain knitted fabric consisting of multi-filament yarn [5]. In order to calculate the porosity of the plain knit structure a formula has been evolved. It has been established that the theoretical values of porosity is almost the same as the experimental one. The porosity increases with the increase of stitch length as pointed out by the experimental as well as theoretical results. The porosity of weft knit structure has been determined with air permeability, image processing, and with the proposed model [6]. The porosity model proposed was for plain knitted fabric made out of mono-filament yarn. It has been concluded that the geometric model is the most suitable and easiest method to determine porosity and it is more influenced by the loop length than the stitch density and the thickness. Moreover, the yarn count or number also influences the porosity.

10.2 Technical details

The pores and the capillaries created by the individual filaments in the spacer yarns have been analyzed geometrically. Considering the spacer structure geometry, the model of the capillary radius has been evolved. The factors considered in the model are fabric thickness, fabric structure, course spacing, wale spacing, yarn crimp, the number of filaments in the yarn, and the yarn counts. Depending on the weight of the fabric, experimental porosities of various spacer structures have been estimated. In order to confirm the validity of assumptions made in calculation of porosity and capillary radius the theoretical porosity has been compared with the experimental porosity.

10.3   Model of geometry of weft knitted spacer structure

The spacer yarn in the fabric is assumed to lie in a zig zag path between front and back bed for the calculation of capillary radius (Figure 1)

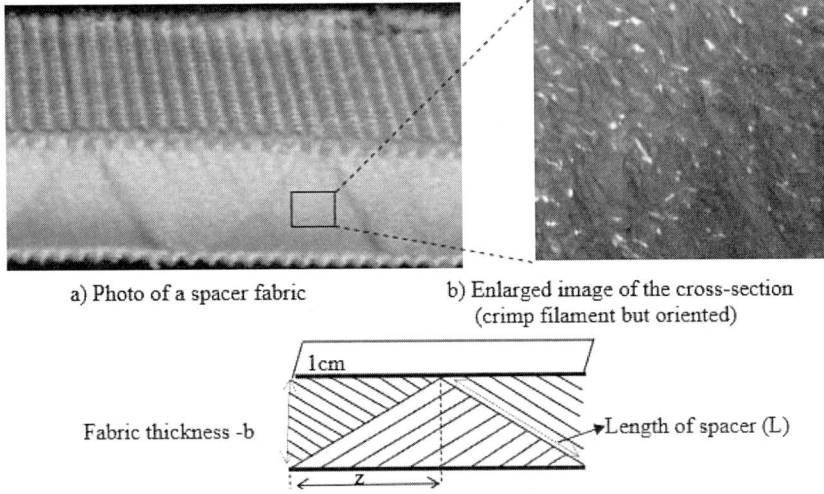

a) Photo of a spacer fabric   b) Enlarged image of the cross-section
                                           (crimp filament but oriented)

c) Diagram of spacer yarns packing in the fabric in course direction

Figure 1 Course direction cross- section of knitted spacer fabric [7]

Since the spacer yarn is untwisted in the spacer structure, it is also assumed that the filaments are equally distributed and parallel to the yarn direction. Further, it can be assumed that the air space in the spacer structure has very fine circular capillaries that lie parallel to the spacer yarn with the length equal to the spacer yarn. Hence the assumption of the capillary model can be given as follows:

a)   The spacer yarn is straight in the structure between front to back bed.

b)   The filaments are crimped and not straight inside the textured yarn.

c)   Since no external compressing forces exist, there is equal distribution of filaments within the yarns.

d)   Owing to the very close packing of the entire spacer structure the spaces between filaments within and between the yarns are very similar .

A cuboid sector of fabric has been considered for model and has the surface with a rectangle (Figure 1.c). Considering the cuboid to have a length 1cm, the course wise distance its width that has one spacer yarn extending the full length between front and back bed, and fabric thickness is the depth. As depicted in Figure 1c, when the number of spaces displaced from one tuck in the front bed to the next tuck in the back bed (floats) is N, then the yarn makes an N x N zigzag layout(Figure 1c).

Considering the fabric thickness as 't'and the distance between the two wales as 'w', where two adjacent tucks are formed from a yarn, joining the front and back beds.

The width of a cuboid, Wis given as

W = N x wale width = N/w,

w – wales/cm

10.5   Determination of porosity and capillary radius in the spacer fabric

The following points need to be considered in measuring the porosity and capillarity of the weft knitted structure

a)   The volume of the cuboid of the test fabric can be considered to have a certain value.

b)   Volume = Thickness x width x length of the fabric

c)   The averages of the fabric thicknesses in wet and dry states are calculated after measuring under a pressure of 1kPa.

d)   Fabric thickness is measured by placing test fabrics between two plates under a pressure of 500g/50cm2 (1kPa).

e)   The variations between dry and wet fabrics are very small when test fabrics are properly relaxed.

f)   The length of spacer yarn between two tucks can then be determined as follows

$$L = \sqrt{(t2+w2)} = \sqrt{[t2 + (S/w)2]}$$

Considering that the fibre volume (V) is the volume of spacer yarns in the cuboid and the weight (ws) is the weight of spacer yarns in the cuboid.

Volume of material (Vm) = weight of yarn (fibre) / fibre density

Weight of material (ws) = Total yarn lengthx yarn count

The total yarn volume within the cuboid has to be calculated differently taking into account the two single jersey courses. The yarn volume of two single jersey courses of the cuboid can be calculated.

This fibre volume of single jersey courses must be added to the volume of the material, when calculating porosity.

Thus the porosity can be given as;

The air space within the considered cuboid = Total volume- volume of the material

Generally, the number of capillaries in a yarn is equal to the number of filaments in the yarn minus the number of filaments on the surface of the yarn, in the case of cylindrical packing.

But, with closely packed spacer structures, filaments on the surface form capillaries with the surface filaments from adjacent yarns. Hence the total number of capillaries is equal to the total number of filaments in the yarn and then the total volume of capillaries is equal to the air space in the fabric occupied by the yarn.

The air space = Total number of capillaries x length x area of one capillary

Then the total number of individual filaments in the yarn is 170 as five fold 1/167/34 polyester yarn has been used as the spacer fibre. The total number of capillaries within the yarn is also considered to be 170, and hence the total number of capillaries within the unit is equal to 170 x rows x c .

The crimp of the filament has not been considered in the calculations. Textured yarns with high crimp have been used as spacer yarns, and also the yarns are in a relaxed state. Thus, the crimp to the material length has to be considered, and the formula for capillary radius becomes:

10.7   Practical assessment of Porosity

The experimental evaluation of the porosity has been conducted by measuring the dry fabric weight, and by calculating the bulk density from the length, the width and the thickness of the knitted specimen. The porosity of the specimen has been determined by a relationship:

In the experimental study of porosity, various kinds of spacer knitted fabrics have been produced.Spacer structures have been produced through variation of spacer yarns in between two single jersey courses and the number of spaces between two consecutive tucks so as to obtain varied porosity, capillary radii and angle of capillaries to the horizontal surface shows an example for one spacer fabric, which has 7 spacer yarns (rows) in-between two single jersey courses and 7 (S) needle spaces in-between two consecutive tucks. Eleven fabrics have been considered which are properly compact and exhibit no differences between intra and inter filament spaces in the cross-sectional view observed in microscope.

Fabric thickness, fabric weight and other fabric parameters like wales/ cm and courses/ cm have been determined.

10.8   Discussion of the findings

Eleven types of weft knit spacer fabrics have been used for the calculation of the theoretical porosity and the measurement of the experimental porosity. The theoretical porosity has been calculated using formula and other fabric parameters. The yarn crimp has been determined after revealing spacer yarn from the structures. The yarn count was taken as 85.3 tex for 5 fold filament yarn, being the specification given by the manufacture. Fibre density was taken as 1.39 as the yarn is polyester.

The experimental porosities are calculated using weight and the volume of the test fabric. The findings reveal that the theoretical values and the experimental values are nearly similar. The theoretical values are only about 0.02 higher than the experimental values. That means the porosity models that have been developed can be used to calculate the porosity of weft knitted spacer structures. Generally, the findings reveal that there is reduction of porosity with the increase of number of spacer yarns. But, the fabric parameters like courses and wales/cm, and the tightness of the spacer structure influence the thickness, which has a great influence on porosity.

The theoretical model developed to calculate capillary radii should be correct since similar assumptions have been made and followed similar steps as in developing the porosity model, which has been proved as valid. The findings show that the fabric parameters like number of spacer courses within a repeat, floating length and stitch density influence the radius of capillaries. Structures having low stitch density and greater number of spacer yarns have finer capillaries.

10.9   References

[1]   J. Chamberlain, "Hosiery Yarns and Fabrics", J.W.Hemming &Capey, Leicester, 1926.

[2]   F. T. Pierce, Textile Research Journal, 27, 123 (1947).

[3]   W. E. Shinn, Textile Research Journal, 25, 270 (1955).

[4]   R. Postle, Journal of Textile Institute, 62, 219 (1971).

[5]   G. B. Delkumburewatte, and Dias, T., in "85th Textile Institute world conferance", Colombo, Sri Lanka, 182,(2007).

[6]   S. Benltoufa, Fayala, F., Cheikhrouhou, M. and Nasrallah, S., AUTEX Research Journal, 7, 63 (2007).

[7]   Dias T, and Delkumburewatte, Porosity and capillarity of weft knitted spacer structures, Fibers and Polymers, April 2009.

Chapter 11

Evaluation of knit spacer fabrics for sound absorption behavior

Summary

Investigation has been carried out on the sound absorption behavior of knitted spacer fabrics. Evaluation and comparison has been done relating to the influences of various fabric layers and arrangement sequences on the noise absorption coefficient. A number of interesting findings emerged from the investigation. On the one hand the weft-knitted spacer fabric show the typical sound absorption behavior of porous absorber, and on the other hand the warp-knitted spacer fabrics show the typical sound absorption behavior of microperforated panel (MPP) absorber. As the frequency of the two types of fabrics increase there is increase in noise absorption coefficient (NAC). Frequency selected sound absorption having a resonance form is observed in warp knit as well as weft knit spacer fabrics. There is enhancement of sound absorption by lamination of various fabric layers. The increase in NACs is considerable in the case of weft-knitted spacer fabric, from the first to fourth layer and further layers are found to be ineffective. Whereas, in the case of warp knitted spacer fabric the NACs there is continuous enhancement as the fabric layers increase, under the condition that the resonance area shifts towards the lower frequency side. The sound absorbance of the combinations of warp and weft knit spacer fabrics can show considerable enhancement, but the sequence of their arrangement has a definite influence. Weft knit spacer fabrics backed with warp knit spacer fabrics exhibit lower NACs compared with the warp-knitted spacer fabrics backed with weft knitted fabrics, at greater frequencies. But, the converse holds good at lower frequencies. In order to obtain high NACs at low and middle frequencies the air-back cavity can be replaced by multilayered warp-knitted spacer fabrics.

11.1 Introduction

Noise has become a serious health hazard and is therefore a matter of environmental concern in day to day human life [1]. There exist a good number of possibilities to reduce noise and are classified as passive and active mediums [2]. In the case of passive mediums noise is reduced by transforming energy into heat. In the case of active mediums the reduction of noise is effected through the application of external energy [3]. The most commonly used passive medium for sound absorption is porous material. Sound absorbability of this kind of material depends on the sound wave frequency. During high frequencies, as the sound waves pass over the uneven pores of porous material, an adiabatic process occurs that creates friction leading to loss of heat [4]. But, at low frequencies the process is isothermal. In such condition, owing to energy loss resulting from heat exchange poroelastic materials absorbs sound [5]. Normally, high frequencies restrict poroelastic efficiency. In order to reduce noise a resonator serves as a type passive medium through conversion of the resonator itself into a vibration. Owing to variations in sound pressure sound is absorbed by a noise resonator through heat loss. In a number of noise control applications perforated panels along with air-back cavity and rigid backing have been used [6]. Microperforated panels (MPPs) having hole diameter less than a millimeter (0.5–1 mm) are considered as one of the best options for third generation sound absorbing materials [7]. During recent times there has been a great focus on textile materials like nonwoven, woven, and knitted fabrics considered as porous material for sound absorption application owing to their economical and ecological merits [8–10]. Many studies have been carried out on nonwoven fiber webs with regard to the noise-absorption properties and theoretical evaluation [11]. But, practical problems exist in designing a textured surface having an attractive appearance for nonwovens eventhough they have desirable noise absorption properties and low cost. Thus nonwoven fiber webs are usually draped with a woven fabric [12]. Plain weft-knitted fabrics have also been considered in sound absorption, but their noise absorption performance is poor [10]. Spacer structures have been introduced for improving the noise absorption ability of knitted fabrics [13, 14]. Investigations have been directed towards the sound absorption properties and theoretical modeling of weft knitted spacer fabrics that comprise of two plain knit surface layers and a spacer layer consisting of multifilament yarns through tuck stitches. Only above 2000 Hz it has been found that the sound absorbency of the weft-knitted spacer fabric is effective. Also, investigations have been conducted on the sound absorbency of the weft-knitted spacer fabric made of monofilament yarn as a spacer yarn along with an uniform pattern of micropores on the surfaces. It has been found that such fabric can enable considerable absorbability at mid-high frequencies, but with a narrower absorption frequency range. Knit spacer fabrics have designable appearances and structures which improve their value addition despite being more expensive compared to nonwovens [15]. The chapter highlights on the study of the sound absorption behavior of both weft and warp-knitted spacer fabrics and their combinations. It has been aimed to identify a sound absorber having wider frequency range and improved sound absorbency at lower frequencies.

11.2   Technical details

Warp and weft knitted spacer fabrics and their combinations have been used in the investigation. In the case of weft-knitted spacer fabric nylon/spandex yarn has been used as the outer layer and textured polyester multifilament used as the spacer yarn. The structure of the weft knit spacer fabric comprise of the top as well as bottom layers having varied plain knitted structure and they are interconnected together with six separate spacer yarns through tuck stitches [16]. After steaming treatment many void pores are formed between two outer layers due to the interconnection of textured polyester multifilament yarns, and hence can be considered as a type of porous sound absorber. Also, the slits formed between the adjacent spacer yarns enable sound waves to penetrate well into the fabric.

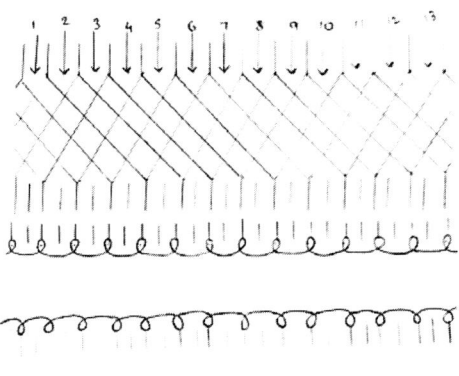

Space yarn.

Figure 1 - Notation of yarn path of weft knit spacer fabric [16]

Raschel knitting machine has been used to produce the warp-knitted spacer fabric. Different chain notations have been used (Figure 1). Unlike in the case of weft-knitted spacer fabric, monofilament has been used as the spacer yarn in the warp spacer fabric. After dyeing, the holes in the elliptical form are evenly distributed on the surface of the fabric. The perforated ratio is the ratio between the surface area of the holes and the total area. The fabric can be considered as a microperforated panel absorber, since the minor axis of each hole is smaller than one mm. Besides, the void spacer between the two outer layers could be considered as an air gap. Fabrics have been conditioned in standard atmosphere. Air permeability tester has been used to measure the air resistance and compression tester used to measure the fabric thickness. The noise absorption coefficient has been measured using microphone impedance measurement tube in order to conduct sound absorbency tests were carried out by measuring the noise absorption coefficient as per prescribed standards. Tube of specific diameter has been designed to measure the noise absorptions of normal incident sounds within 0–6400 Hz. The test fabrics have been cut to circular shape with specified dimension. They have then been placed into the impedance tube either being close to the metal plunger or leaving a distance as air-back cavity based upon conditions of experiment.

In order to test multilayered fabrics, the fabrics should be closed and conjoined together. The single layer and multilayer weft-knitted and warp-knitted spacer fabrics that have been laminated with the same or different kinds of fabrics, have been tested in this investigation.

11.3   Analysis of sound absorption behavior of single layer spacer fabrics

Figure 2 depicts the noise absorption coefficients (NACs) of single layer spacer fabrics having no air-back layer. As the frequency increases, the NACs of both fabrics X and Y increase. Fabric X shows NAC values between frequency ranges that typify sound absorption behavior of porous material. Fabric Y exhibits lower NACs than fabric X for all the frequencies. Moreover, fabric Y shows curve with two slight peaks. It exhibits the typical behavior of MPP absorbers resulting from combination of various sizes of perforated holes [16]. Because of the presence of the slits resembling perforated holes, fabric X can also be considered as a perforated panel absorber in the presence of the air-back cavity. Figure 6 depicts the NACs of fabrics having two different thicknesses (8 and 16 cm) of air-back cavity. All the sound absorption curves appear as frequency spectra, since the existence of the air-back cavity creates frequency-selected sound absorptions owing to the system resonance. Further, the presence of the air-back cavity leads to narrower absorption frequency ranges, but with higher NACs at lower frequencies than that where no air-back cavity is used. The NACs of the fabrics are also affected by the thickness of the air-back cavity. The increase of the air-back cavity makes the absorption frequency move toward the lower frequency side.

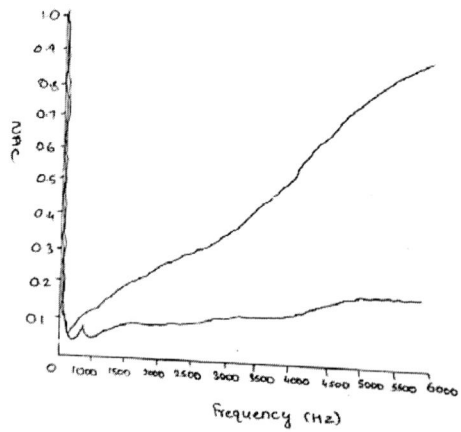

Figure 2 – NACs of single layer spacer fabrics without air back cavity[16]

Figure 2 and 3 also show that one test fabric exhibits better sound absorbability than the other for both cases with and without the air-back cavity because of greater thickness. But, its sound absorbency at lower frequencies is still very low without the use of air-back cavity.

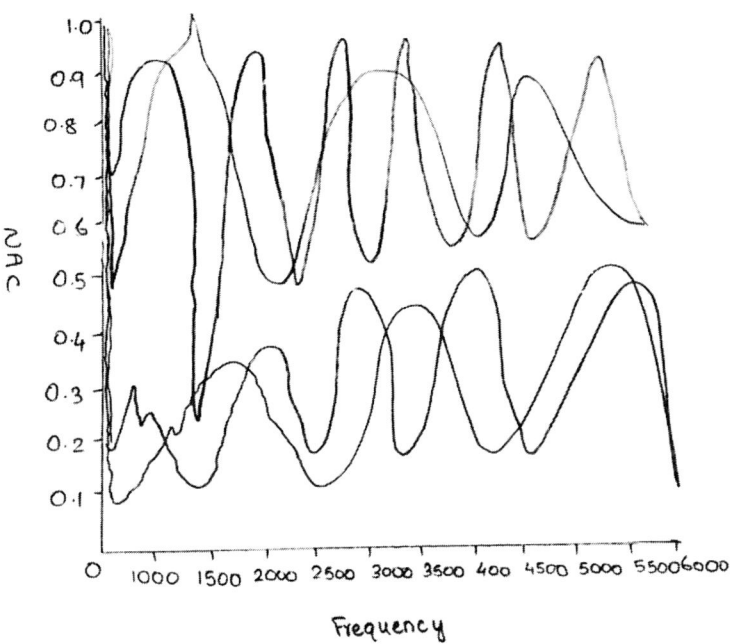

Figure 3 – NACs of spacer fabrics with air back cavity[16]

11.4 Sound absorption behavior of multilayered spacer fabrics

From the aforesaid discussions it can be seen that the single layer spacer fabric have low NACs at lower frequencies. The frequency range has been very narrow despite the use of the air-back cavity improving the NACs. Also, the use of the air-back cavity is suitable for the practical use of spacer fabrics. For enhancing the NACs at lower frequencies, the use of laminated absorbers having increased thickness is the generally used technique. The NACs of the sound absorbers laminated with the similar type of spacer fabric have been studied. In the case of porous sound absorbers, the NACs are based on their thickness, porosity, airflow resistivity and appearance [16]. Broadly stated, the thickness increase can significantly improve the NACs at low frequencies. But, the thickness increase does not have the evident influences on NACs at high frequencies and can occasionally result in slight decrease in NAC. In addition, the influence of increasing the thickness for improvement of the NACs at low frequencies is restricted, as the NACs do not increase further after the thickness attains a critical value. Figure 4 proves this phenomenon can be confirmed, where the NACs of the porous sound absorbers laminated together having various layers of the similar weft-knitted spacer fabric have been depicted. It is very clear that the two-layered spacer fabrics already show enhancement in the sound absorption capacity of the spacer fabric at low frequencies. But, there is decrease in improvement after two layers, and the influence of increasing the thickness on the NACs cannot be seen further after five layers. In order to further enhance the NACs, other methods like increase of the porosity, airflow resistivity and fabric surface smoothness or a combination of various types of fabrics, can be taken into account.

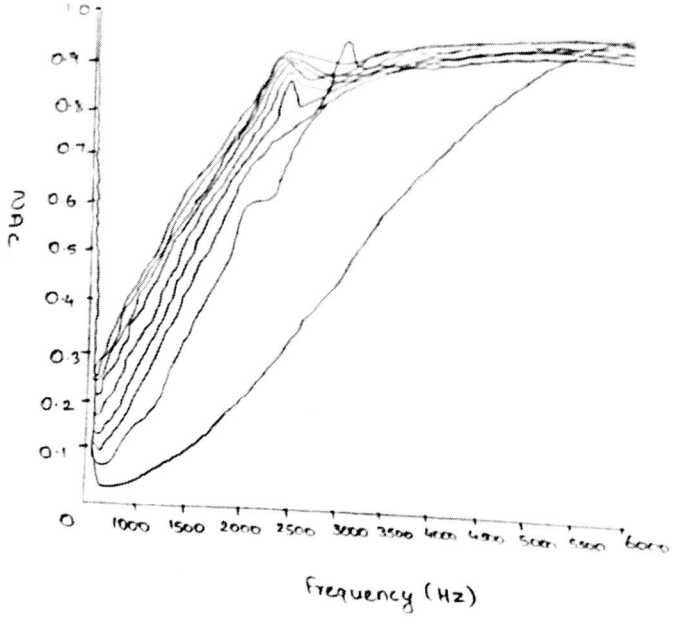

Figure 4 – NACs of weft knit spacer fabrics laminated with different layers[16]

As shown in Figure 5, the peaks around 1600 Hz for the absorbers laminated from 2-8 layers of spacer fabrics can be the resonance of the system. In the case of MPP absorbers, the performance ratio determined the NACs, dimension of the perforated holes, panel thickness and thickness of the air-back cavity. The MPP absorbers provide high absorbability at mid-high frequencies. But, usually their absorption frequency ranges are restricted due to their nature as a resonator. In order to achieve a broader frequency range for sound absorption, the combination of multilayered MPP absorbers with various frequency characteristics or the use of a MPP absorber having various size of perforated holes are always used.

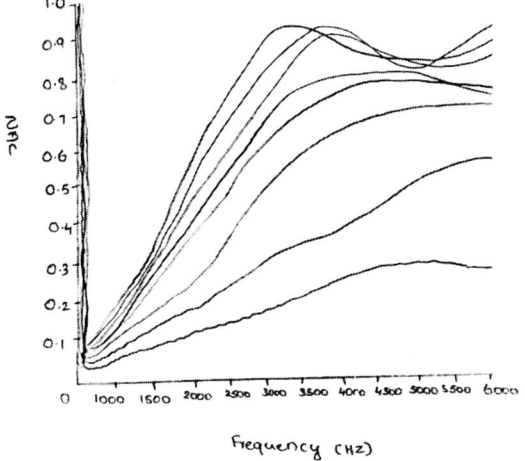

Frequency (Hz)

Figure 5 – NACs of warp knit spacer fabrics laminated with different layers[16]

Figure 5 depicts the NACs of the MPP absorbers formed with various layers of warp-knitted spacer fabrics. As the fabric layers increase, the NACs increase considerably. Also, the resonance phenomena can be seen for all the fabric layers and their resonance region moves towards the lower frequency side on increasing the fabric layers. On analysis of the curve of the thickest absorbers formed with 8 layers of a fabric, two peaks respectively at 2128 Hz and 6184 Hz are found because of resonance. As per the curves in Figure 5, it can be predicted that as the layers of fabrics increase there is a continuous increase in NAC. Hence, an absorber having better sound absorbability can be obtained when sufficient layers are used. As discussed previously, such behavior is different from that of the sound absorbers formed with various layers of weft-knitted spacer fabric. Depending on the sound absorption principle of MPPs, various types of warp-knitted spacer fabrics can be designed having various frequency characteristics. The lamination containing various warp-knitted spacer fabrics or the use of various sizes of meshes size on the same fabric can be potential ways to obtain a wider range of absorption frequency.

11.5 Sound absorption behavior of combined multilayered spacer fabrics

For industrial uses, multilayered sound absorbers consisting perforated panels as well as porous materials have been well utilized for broadband noise absorptions [15]. Hence, the sound absorption behavior of multilayered spacer fabrics laminated with various combinations of fabric with porous material and perforated panel have been investigated. There are two cases considered wherein fabric A has been placed on the front or the back of fabric B. Figure 6 depicts the NACs of various layers of fabric B backed with one layer of fabric A. When compared with previous figure, the sound absorbability of the warp knit spacer fabrics can significantly be enhanced by backing one layer of fabric A [16]. Also with the increase in the layers of fabric B, the resonance regions are found to shift towards the lower frequency side. Figure 8 already depicts such a trend. But, the NACs for the low frequencies under 500Hz are still less than 0.5, which is a critical value normally used to find out the usefulness of a sound absorber. The NACs of various layers of fabric B backed with double layers of fabric A have also been studied, as double-layered weft-knitted spacer fabrics already have a very obvious improvement of the sound absorption capacity at lower frequencies (Figure 3).

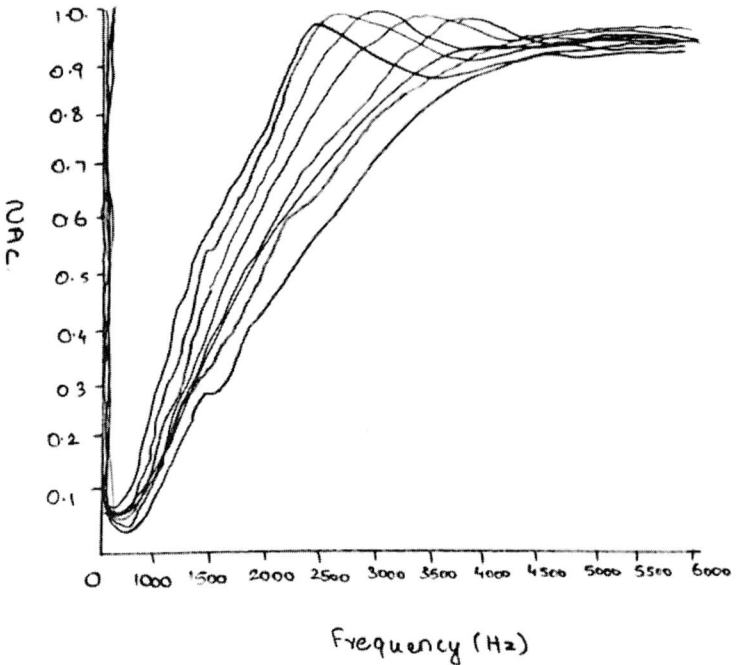

Figure 6 – NACs of different layers of warp knit spacer fabric backed with one layer of weft knit spacer fabric[16]

Figure 7 depicts the findings. In such a situation the NACs at lower as well as higher frequencies are further improved and a broader absorption frequency range is obtained. It can also be observed that the NAC of fabric (8B-2A) attains a value of 0.5 at 500Hz. The curves in Figure 6 show that the curved wave forms at the middle as well as high frequencies become straighter, and hence the sound absorption behavior of the fabrics becomes more stable. With the further increase in number of layers of fabric A, the NACs at low frequencies can further be improved. However, this improvement should be limited as the results shown in Figure 8 have already confirmed that there is no further increase in the NACs as the thickness attains a critical value. On the other hand, increasing the layers of the fabrics will lead to an increase in cost and consumption of material. Also, when the practical applications are considered, far high thickness of the sound absorbers are not found to be suitable. The sound absorption behavior is altered to an extent by changing the arrangement sequence, i.e., by swapping the positions of weft-knitted and warp-knitted spacer fabrics.

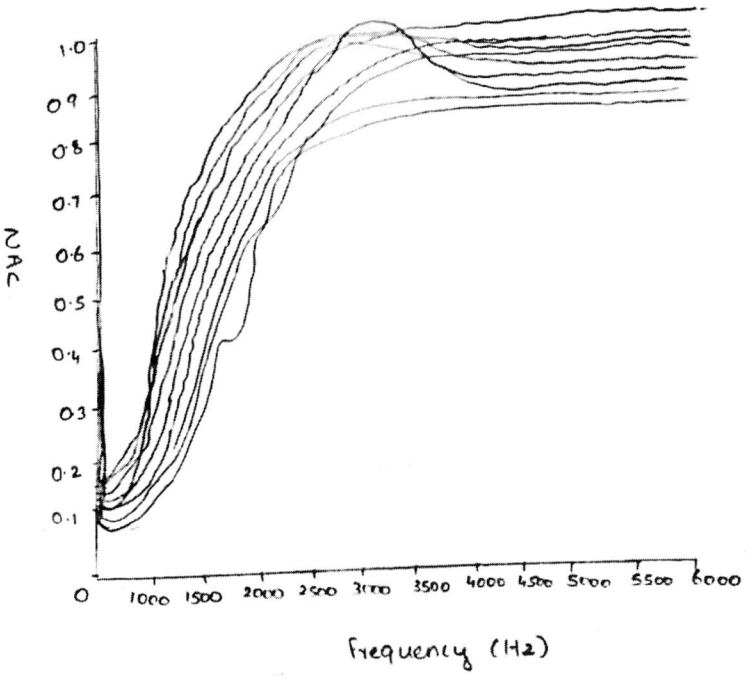

Figure 7 – NACs of different layers of warp knitted spacer fabric backed with two layers of weft knit spacer fabric[16]

Figure 7 depicts the NACs of one layer of A backed with various layers of B. In comparison with Figure 5, the sound absorption capacity of the multilayered spacer fabrics at low frequencies show much better improvement due to increase in number of layers in fabric B. However, at higher frequencies the NACs are far lesser, though their values are above 0.5. The similar phenomenon can also be noticed, i.e., the shift of the resonance regions towards the lower frequency side. The resonance gives rise to small peaks around 1500 Hz. Further improvement in the NACs occurs at low frequencies. Also, the curved wave forms for the middle and high frequencies become straighter. Under such a situation the small peaks around 1500Hz caused by resonance are still noticed.

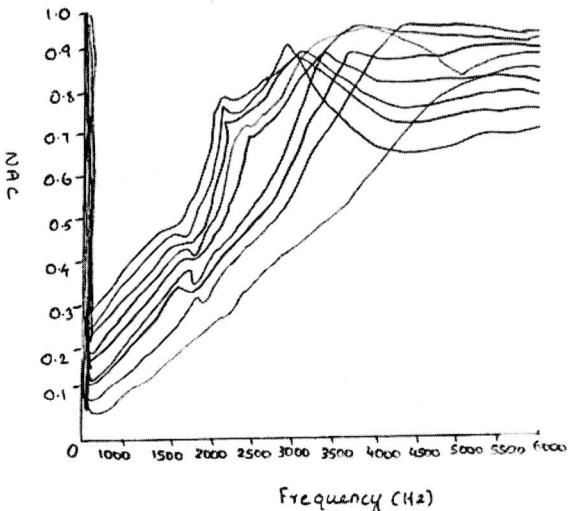

Figure 8 – NACs of one layer of weft knit spacer fabric backed with different layers of warp knit spacer fabric[16]

The aforesaid discussions reveal that the arrangement sequence has an evident influence on the sound absorption. It arises due to the various positions of fabric B which lead to various sound absorption effects. When fabric B is placed before fabric A, the fabric B performs as a resonator absorber, and proves beneficial in improving the NACs for all the frequencies. Whereas, as fabric B is placed behind fabric A, it functions as a air-back cavity and results in an increase of the thickness of the system. Under such situation, the NACs of the system can be considerably enhanced at low frequencies, but with a cost of NAC reduction at high frequencies.

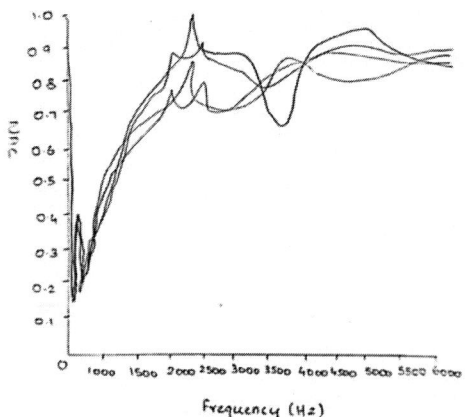

Figure 9 – Comparison of the NACs for weft knit spacer fabric backed with air layer or eight layered warp knit spacer fabric [16]

Figure 9 depicts the NACs of one or two layers of fabric A respectively backed with 4 cm air layer and eight layers of fabric. In this case, the thickness of the air-back cavity is near to that of eight layers of fabric B. The curves between the same layered fabric A backed with the airback cavity and eight layered fabric B are almost coincident when the frequencies are lower than 3000 Hz. The differences of the NACs above 3000 Hz arise from the minor difference in thickness between eight layers of fabric B and the air-back cavity as well as the area occupied by fabric B. Such findings show that air-back cavity can be replaced by multilayered warp-knitted spacer fabrics to attain high NACs at low and middle frequencies. It can be a suitable method for the practical end use.

11.6   References

[1]   Lou, C. W., Lin, J. H., and Su, K. H., Recycling Polyester and Polypropylene Nonwoven Selvages to Produce Functional Sound Absorption Composites, *Textile Res. J.* **75**(5), 390–394 (2005).

[2]   Sagartzazu, X., Hervella-Nieto, L., and Pagalday, J. M., Review in Sound Absorbing Materials, *Arch. Comput. Methods Eng.* **15**(3), 311–342 (2008).

[3]   Kuo, S. M., Morgan, D. R., "Active Noise Control Systems," Wiley, New York, 1996.

[4]   Cummings, A., and Chang, I. J., Acoustic propagation in porous media with internal mean flow, *J. Sound Vib.* **114**(3), 565–581 (1987).

[5]   Zhang, B., and Chen, T. N., Calculation of sound absorption characteristics of porous sintered fiber metal, *Appl. Acoust.* **70**(2), 337–346 (2009).

[6]   Sakagami, K., Morimoto, M., and Yairi, M., A Note on the Relationship between the Sound Absorption by Microperforated Panels and Panel/Membrane-type Absorbers, *Appl. Acoust.* **70**(8), 1131–1136 (2009).

[7]   Maa, D. Y., Potential of microperforated panel absorber, *J. Acoust. Soc. Am.* **104**(5), 2861–2866 (1998).

[8]   Tascan, M., and Vaughn, E. A., Effects of Total Surface Area and Fabric Density on the Acoustical Behavior of Needlepunched Nonwoven Fabrics, *Textile Res. J.* **78**(4), 289–296 (2008).

[9]   Shoshani, Y., and Rosenhouse, G., Noise absorption by woven fabrics, *Appl. Acoust.* **30**(4), 321–333 (1990).

[10]   Dias, T., and Monaragala, R., Sound Absorbtion in Knitted Structures for Interior Noise Reduction in Automobiles, *Meas. Sci. Technol.* **17**(9), 2499–2505 (2006).

[11]   Shoshani, Y., and Yakubov, Y., A Model for Calculating the Noise Absorption Capacity of Nonwoven Fiber Webs, *TextileRes. J.* **69**(7), 519–526 (1999).

[12]   Shoshani, Y. K., Noise Absorption by a Combination of Woven and Nonwoven Fabrics, *J. Text. Inst.* **82**(4), 500–503 (1991).

[13]   Dias, T., Monaragala, R., and Lay, E., Analysis of Thick Spacer Fabrics to Reduce Automobile Interior Noise, *Meas. Sci. Technol.* **18**(7), 1979–91 (2007).

[14]   Dias, T., Monaragala, R., Needham, P., and Lay, E., Analysis of Sound Absorption of Tuck Spacer Fabrics to Reduce Automotive Noise, *Meas. Sci. Technol.* **18**(8), 2657–2666 (2007).

[15]   Zhu C. Y., and Huang, Q. B., A Method for Calculating the Absorption Coefficient of A Multi-layer Absorbent Using the Electro-acoustic Analogy, *Appl. Acoust.* **66**(7), 879–887 (2005).

[16]   Yanping L and Hong H, Sound Absorption Behavior of Knitted Spacer Fabrics, Textile research journal, Vol. 80(18): 1949.

Chapter 12

Mechanical and physical properties of 3D knit spacer fabrics

Summary

Quantitative study has been carried out on spacer fabrics with regard to different fabric characteristics that include air permeability, thermal conductivity and low-stress mechanical properties. The kind of spacer yarn and its arrangement greatly influence the compression properties. The type of fabric, structure, type of spacer yarn and density bears a close relationship to the bending properties whereas the type of spacer yarn and fabric strongly influence the stretch and recovery properties of the spacer fabrics. The air permeability, thermal conductivity and mechanical properties of spacer fabric could possibly be strongly influenced by the spacer fabric characteristics. Hence it becomes crucial to properly choose spacer fabric based on its required area of application.

12.1 Introduction

A spacer fabric is basically a 3D knit fabric having two different fabric layers that are connected together or separated by spacer yarns [1-3]. The spacer fabric may be of warp knit or weft knit categories. Raschel knitting machine is used to knit the warp knit spacer fabrics while circular knitting machine is used to knit the weft knit spacer fabrics [4-7]. The areas of applications of spacer fabrics are many and include mobile textiles (car seat covers, dashboard cover), industrial textiles (composites), medical textiles (anti-decubitus blankets), sports textiles and foundation garments (bra cups, pads for swimwear) [8-10]. Spacer fabric as a component material is highly breathable, thus creating a moisture free environment, which in turn reduces the chances of skin maceration. They result in a more enhanced degree of comfort than materials like foam, neoprene and laminate fabrics. Contrary to polyurethane foam spacer fabrics can be considered eco friendly textile materials, as they are recyclable [2,8]. For over years the spacer fabrics have been investigated across the world [2,3]. But, the influence of spacer fabric characteristics on its physical and mechanical properties have not been explored much. The chapter highlights the effect of a number of fabric structures on the physical and mechanical properties of spacer fabric.

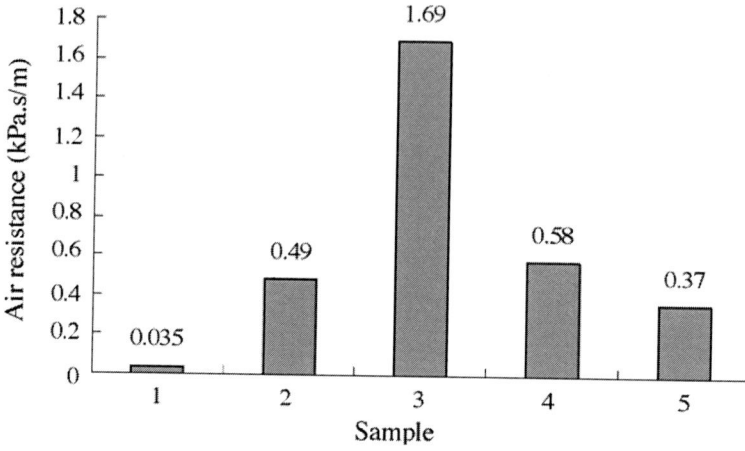

Figure 1 – Comparative air resistance of test fabrics [17]

12.2   Technical details

The investigations involved both warp as well as weft knit spacer fabrics. The fabric characteristics considered are the fabric density, spacer yarn type, thickness of the spacer fabric, spacer yarn diameter and arrangement. The studies have been carried out under standard conditions.

The following tests have been conducted on the fabrics

a)   Air permeability

b)   Thermal conductivity

c)   Bending and compression

d)   Stretchability and recovery

Table 1 – Definitions of bending and compression properties obtained by KES-F system [17]

Properties	Symbol	Definition	Unit
Bending properties			
Bending hysteresis	B	Average slope of the linear regions of the bending hysteresis curve to 1.5 cm^{-1}	uNm
Compression properties			
Compressional resilience	RC	Percentage energy recovery from lateral Compression deformation	%
Fabric thickness at 50gf/cm^2	T_m	Fabric thickness at 50gf/cm^2	mm

The fabric characteristics of various spacer fabrics have been studied. Normally, the knit spacer fabrics have a thickness between 1.5 mm to 60mm. In the case of sports textile and foundation garment the fabric thickness used for investigation range between 2.8mm to 4mm. Based on the thickness of the spacer fabric structure and the type of joining yarns its compression resistance can be varied.

Table 2 – Structural details of test fabrics [17]

	Sample 1 (WA-MO)	Sample 2 (WE-MO-1)	Sample 3 (WE-MU-1)	Sample 4 (WE-MU-2)	Sample 5 (WE-MO-2)
Fabric Structure					
Front view					
Back view					
Side view - weftwise					
Side view - warpwise					

Monofilament as well as multifilament joining yarns of different diameters has been used. Even though the spacer yarns have been positioned at right angle to the two outer fabrics, it is very much possible thaton application of pressure, the yarns would simply be pushed sideways, and hence the compression resistancedecreases [3]. Hence, the spacer yarns have subsequently been arranged in a V-shaped configuration and it is possible to calculate the spacer yarn arrangement angle considering spacer fabric thickness and segment width.

12.3   Air permeability and thermal conductivity

Figure 1 depicts the air resistance of various spacer fabrics. As the number of kPa s/m increases the air resistance of the fabric increases [11]. The thermal conductivity of different spacer fabrics has also been recorded. The heat transfer from skin to the fabric surface becomes faster arises from a higher value of thermal conductivity, it results in a cooler feeling. The thermal conductivity of various knit spacer fabrics is depicted in figure 2.

The test fabrics showed the lowest air resistance and thermal conductivity as well as the highest. The construction characteristics of the yarns in the spacer fabrics strongly influence their air permeability, and their large volumes are occupied by air. The air permeability of the spacer fabric is influenced by a number of parameters, which include fabric structure, thickness, surface characteristics, etc. [12]. The thermal property and air permeability of the spacer fabric are most significantly influenced by the fabric density [17]. A higher fabric density will hinder the air flows through the fabric, thus resulting in a poor air permeability of the fabrics. But, since a higher fabric density contains less space to trap air inside, it will have a better thermal conductivity, as there will be less space to trap air inside. Hence a denser fabric will exhibit better thermal ventilation.

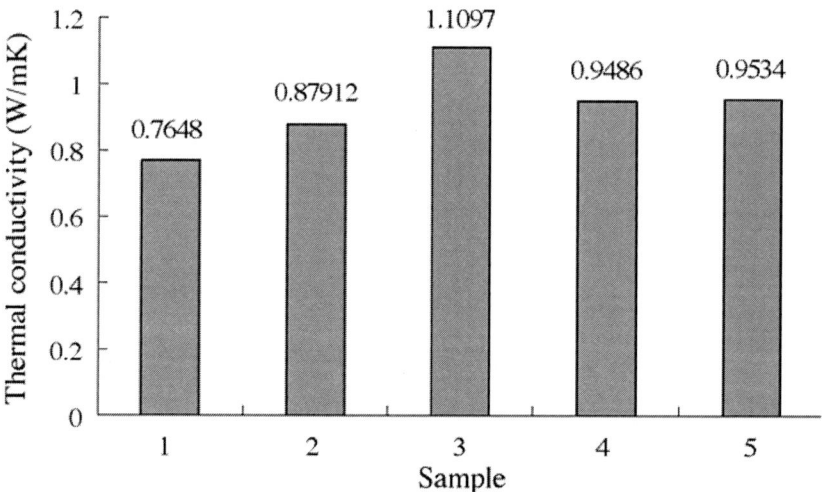

Figure 2 – Thermal conductivity of different test fabrics [17]

12.4 Compression properties

Figure 3 depicts the compression resistance of various spacer fabrics under specified pressure with regard to the percentage change in thickness. A lower compression resistance results from a higher percentage of thickness compressed. Fabrics with multifilament spacer yarn are found to have lesser compression resistance than those made from monofilament spacer yarn. Comparative studies on the spacer fabrics with identical type of spacer yarn have revealed that the compression resistance of a fabric bears a close relation with the spacer yarn arrangement. The fabric having a larger angle of inclination of spacer yarns will thus exhibit a higher compression resistance, assuming that sane material and diameter of the spacer yarn are used are the same [17]. The compressive resilience is the percentage energy recovery from deformation due to lateral compression. Better recovery property results from a higher percentage of compressive resilience of the fabric. The recovery properties after compression are largely based on the type of spacer yarn.In the case of spacer fabrics with multifilament spacer yarns, the recovery properties are lesser than those with monofilament spacer yarns.

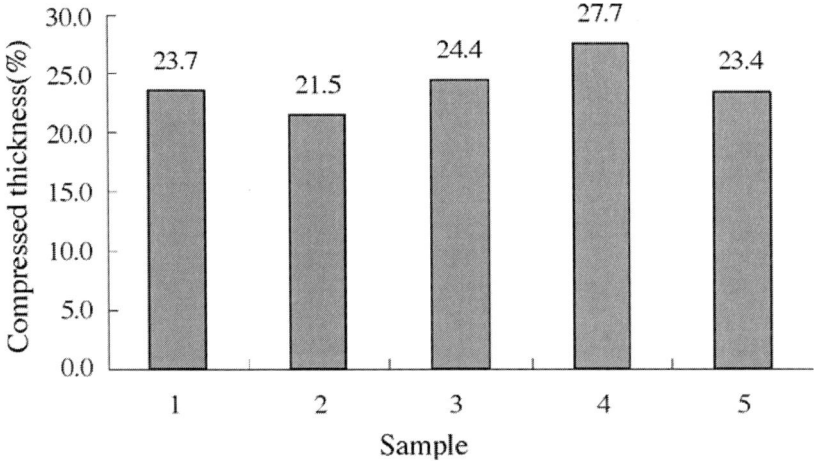

Figure 3 – Compression resistance of test fabrics [17]

12.5 Bending properties

Various spacer fabrics have been analyzed for the bending rigidity. The test results of spacer fabrics in warp as well as weft have been determined [17]. The type of spacer fabric has a close relation with its bending rigidity. Hence, a warp knit spacer fabric has greater bending rigidity in along the warp direction whereas a weft-knit spacer fabric shows a greater bending rigidity along the weft. It could be attributed to the directionality of the incorporated yarn [13,14]. Considering identical kind of fabric(say weft way) it can further be concluded that the bending rigidity is has a close relation to the density of fabric, spacer structure and spacer type [15]. Greater bending rigidity has been found in weft-knitted spacer fabrics with interlock structure, monofilament spacer yarn and a higher fabric density.

12.6 Stretch and recovery

Various spacer fabrics have been investigated for the stretch and recovery properties. The warp knit spacer fabric with 52% polyester and 48% polybutylene terephthalate show the best stretchability in the weft direction and the poorest stretchability in warp direction. The weft knit spacer fabric having 92% polyester and 8% spandex show the best recovery property in both the warp-wise and weft-wise fabric configurations while the warp knit spacer fabric with 52% polyester and 48% polybutylene terephthalate show the poorest recovery property in both directions. Such results indicate that the stretchability of the spacer fabrics relates closely to the kind of fabric [17]. The results depicted in Fig. 5 show that the stretchability of a warp-knitted spacer fabric is high only in the weft-wise direction. On the other hand the stretchability is very less in the warp-wise direction (below 50%). Whereas, weft-knitted spacer fabrics exhibit similar and high stretchability in the weft-wise as well as warp-wise directions. Since the spacer fabric comprises of two separate surface fabrics and are connected together by a spacer yarn, it can thus be stated that spacer fabrics have the same fabric stretchability as their fabric types (i.e. warp-knitted or weft-knitted). When the results of the weft-knitted spacer samples were compared, the stretchabilities of the weft-wise direction of samples 3 and 4 were found to be higher than those of samples 2 and 5. This is due to samples 3 and 4 using multifilament spacer yarns, which have higher stretchability than those corresponding to samples using monofilament spacer yarns [16].

12.7 References

[1]   New patterning possibilities, 2001. Kettenwirk-praxis 2, E14–E15. Raz, D.S., 1993. Flat knitting Technology. Universal Maschienfabrik, Westhausen.

[2]   Wilkens, C., 1993. Raschel knitting spacer fabrics. Kettenwirk-praxis 3, E18–E20.

[3]   Lehmann, W., 1994. Elastic, moulded spacer fabric. Kettenwirk-praxis 3, E19–E20.

[4]   McCartney, P.D., Allen, H.E., Donaghy, J.G., 1999. Underwire brassiere, warp-knitted textile fabric for use in fabricating same, and method of warp knitting such fabric. USPTO Patent Full Text and Image Database, US Patent No. 5669247.

[5]   Donaghy, J.G., Azuero, I.M., 1999. Moldable warp-knitted fabric and method of forming a seamless molded fabric portion there from. USPTO Patent Full Text and Image Database, US Patent No. 5855124.

[6]   Shepherd, A.M., 2004. Weft-knitted spacer fabrics. USPTO Patent Full Text and Image Database, US Patent No. 6779369 B2.

[7]   Willmer, R., 2005. Circular knitting machine, especially for the production of spacer fabric. USPTO Patent Full Text and Image Database, US Patent No. 6915666B2.

[8]   Heide, M., 2001. Spacer fabrics: trends. Kettenwirk-praxis 1, E17–E20.

[9]   Spacer fabrics in medicine, 1999. Kettenwirk-praxis 1, E18–E19. Sytz, R.M., 2004. Knitted stretch spacer material and method of making. USPTO Patent Full Text and Image Database, US Patent No. 6755052B1.

[10]   Bras cups made from a new spacer fabric, 2001. Kettenwirk-praxis 2, E2–E3.

[11]   Yip, J., Chan, K., Sin, K.M., Lau, K.S., 2002. Low temperature plasma treated nylon fabrics. Mater. Process. Technol. 123, 5–12.

[12]   Zhang, P., Gong, R.H., Yanai, Y., Tokura, H., 2002. Effects of clothing material on thermoregulatory responses. Text. Res. J. 72/1, 83–89.

[13]   Machova, K., Klug, P., Waldmann, M., Hoftmann, G., Cherif, C., 2006. Determining of the bending strength of knitted spacer fabric. Melliand Textilberichte 87 (6), E93.

[14]   Reisfeld, A., 1996. Warp Knit Engineering. National Knitted Outerwear Association, New York.

[15]   Kawabata, S., Niwa, M., 1996. Modern Textile Characterization Methods. Marcel Dekker, New York, pp. 329–354, (Chapter 10).

[16]   King, R.R., 1985. Textile Identification, Conservation, and Preservation. Noyes Publications, Park Ridge, N.J., U.S.A.

[17]   Joanne Y, and Sun Pui Ng, 2008, Study of three-dimensional spacer fabrics: Physical and mechanical properties, Journal of Materials Processing Technology 206(1-3):359

Chapter 13

Use of flat knit spacer fabric in light weight composite applications

---

Summary

Innovative type of flat knitting technique has been used to produce newer type of 3D spacer fabrics having individual surface and connecting layers including also the single axis curvatures to the wale direction by means of glass fibre-polypropylene filament hybrid yarns. Up to 4 reinforcement layers of the reinforcement yarns have been integrated into all surface and connecting layers so as to enhance the wall thickness to improve the mechanical performance of the composites. In order to monitor the structural health of end products, the flat-knitting technology has been developed further for the creation of sensor networks through innovative integration of functional yarns into 3D spacer fabric structures during the production of the multilayered 3D spacer fabrics. The high degree of flexibility in the latest technology developed has enabled the easy integration of a broad range of materials into multi-layered knit structures. It can be a break-through development in fully automated production of function-integrated textile performs where such seamless integration of functional yarns into the fabric structures is executed simultaneously, together with forming the multi-layered 3D spacer fabrics.

The innovative 3D spacer fabrics manufactured using new flat knitting technologies through economical single phase production holds good promise in various areas, as for instance, in textile-reinforced concretes, architectural designs, energy sectors, protective textiles, industrial textiles, and geo-textiles taking the scope beyond the potential application trend in the area of lightweight composite structures.

13.1   Introduction

Spacer fabrics are complicated 3D constructions comprising of two different layers of fabric that link vertically with pile yarns or fabric layers. Knitting and weaving techniques have been used to produce the conventional spacer fabrics consisting of two surface layers bound with pile yarns. But, because of inferior mechanical properties, like elasticity and deformability under applied loads, conventional spacer fabrics are unsuitable for high- performance composite applications. Also, the limited distance between the plane layers contributes to the setbacks of these spacer fabrics. A possibility is to connect the planes through vertical fabric layers instead of pile yarns. Such kind of 3D spacer fabric having multi-layer reinforcements in the fabric structures is supposed to exhibit superior mechanical properties and be particularly suitable as textile preforms for lightweight composite applications [1–14]. Future applications of composites made from 3D multi-layer spacer fabrics involve the replacement of conventional panel structures that are being used for aircraft, transport vehicles, marine applications and infrastructures, lift cabins, and ballistic protection for buildings and combat vehicles, etc. But, for the development of these newer types of 3D spacer fabrics, modern electronic flat-knitting machines offer the best option, as they are capable of producing 3D complex-shaped engineering structures. Special technical aspects which permit rapid and complex production include individual needle selection capability, the presence of holding down sinkers, presser-foots, racking, transfer, adapted feeding devices coupled with a computer-aided design (CAD) system, and modern programming installations. Moreover, the flexibility of the knitting process in combination with the possibility of integration of reinforcement yarns into fabric structures has attracted a number of researchers [1–17]. Also, knitted performs show higher drape-ability and greater impact resistance in composites than the commonly used textile structures (e.g. braiding, weaving, unidirectional techniques). Earlier studies have pointed out the analysis and the production of 3D spacer fabrics on the basis of surface and connecting layers, the development of 3D spacer fabrics possessing only the weft inlays3–5 and the development of multi-layered 3D spacer fabrics [3-5,7,8,12]. The research work documented in this direction includes the basic production principles of some flat-knitted spacer fabrics without reinforcements and the theoretical presentation of knitted sandwich spacer fabrics [10,13]. Also, the further advances in flat-knitted multi-layered 3D spacer fabrics including an increased number of reinforcement layers into all individual fabrics are required for high mechanical performance in complex-shaped lightweight composite applications. The sensor networks realized into 3D spacer fabric structures by means of integrating the functional yarns during the single-step production of the multi-layered 3D spacer fabrics can be the break-through development in fully automated production of function-integrated textile performs and these sensor networks could be used to monitor the structural perfection of end products [7,8,10,14–18]. But, owing their aniosotropic mechanical properties and lower chemical and environmental resistances fibres cannot be used alone in structural engineering end uses. Hence, they are embedded in matrix materials to form fibrous composites.

The textile composites can be designed based on the chemical nature of the matrix materials that are normally thermoset and thermoplastic polymers.

Thermoplastic composites contain a minimum of one reinforcement material and a thermoplastic polymer as matrix. Such composites have more merits than thermoset composites. They seem to hold a good potential in some industrial end uses because of their high fracture toughness, easy recycling, elongation, short processing time, different forming possibilities, weld-ability, low cost, and resistance to medias and corrosion [3–5,12,15–17,19]. On the other hand, two stage processes are involved in thermoplastic composite production routes: A precursor material is formed by means high-performance fibers and thermoplastic polymers, followed by the transformation of the component into the final product through application of high pressure and temperature. The commingled hybrid yarns with homogeneously mixed reinforcement and matrix filaments are soft, flexible, drapeable, and are cost effective, which renders them a forerunner for thermoplastic composite applications.

Usually high performance fibres like glass, carbon, and aramid constitute the reinforcement component of the hybrid yarn. Owing to low material cost and better mechanical properties glass fibers are being widely used.

In order to withstand the applied forces better and to ensure good adhesion between the fibres and matrix material and evolve cost effective products particularly suited to the automotive industries the thermoplastic matrix is used to fix the reinforcement components in a specific manner [3,4,12,15–17,19–23]. Glass filaments (GFs) and polypropylene (PP) filaments in hybrid yarn have been blended with a proportion of 52% and 48%, respectively in order to optimize the mechanical properties of textile reinforced thermoplastic composites [17]. Depending on such factors, the commingled hybrid yarn consisting of GFs and PP filaments is preferred for the development of 3D spacer fabrics [3–5,12,14–17]. The objective of the textile-reinforced composite components for function-integrating multi material design in complex lightweight applications is to evolve multilayered 3D spacer fabrics as complex-shaped textile preforms through the flat-knitting method in combination with hybrid yarns for high-performance thermoplastic composite applications. This has been carried out by producing, multi-layer reinforced 3D spacer fabrics curved in various angles (in the warp direction) with exemplar GF-PP filament hybrid yarns by adopting the recently developed flat-knitting methods.

Using such innovative production methods, reinforcement yarns have been integrated up to four layers (two warp and two weft yarns aligned as a multi-layer structure) as biaxial inlays into all surface and connecting layers of spacer fabrics. Functional yarns have also been effectively integrated into spacer fabric structures in a single processing step with the innovative integration concept so as to create sensor networks for structural health monitoring of end products.

13.2 Technical details

Hybrid yarns with GFs (volume: 52%) and PP filaments (volume: 48%) have been used as high-performance and thermoplastic materials, respectively. The modified PP filaments and the GFs have been used as thermoplastic and reinforcing materials. Hybrid yarns of finer linear density have been chosen as the knitting yarns for base fabrics and (base loop yarn) and coarser yarns as reinforcement yarns. Also, carbon filament and copper wire have been integrated as functional yarns into the spacer fabric structures so as to create sensor networks. Electronic knitting machine has been used to knit the spacer fabric. Modern flat knitting technology developed for the purpose has been used to manufacture multi-layer reinforced 3D spacer fabrics with 4 reinforcement layers, into all fabric layers. This innovative flat knitting technology aids in the effortless construction of both plane and connecting fabrics, each being made of four reinforcement layers resulting in increased fabric specific weights (resulting in higher wall thickness of the composite). Figure 1 depicts the technological basis of producing this new type of 3D spacer fabrics.

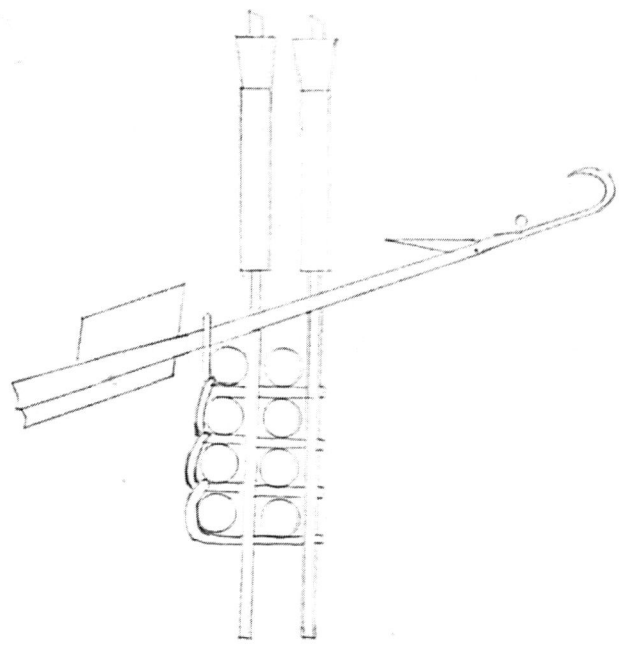

PHASE =

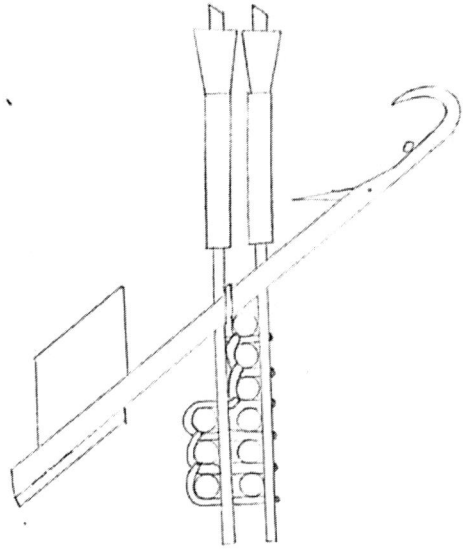

PHASE I

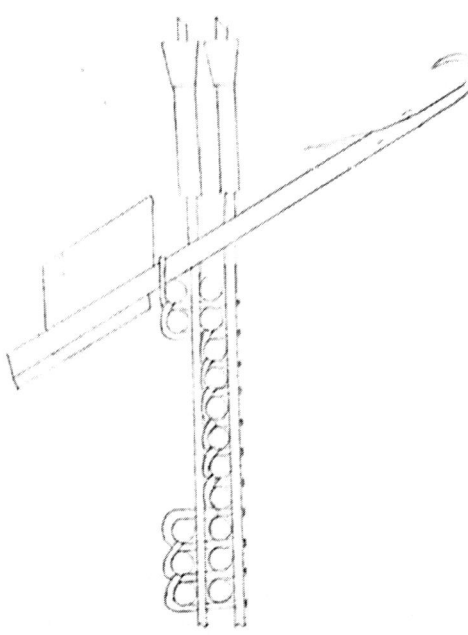

PHASE II

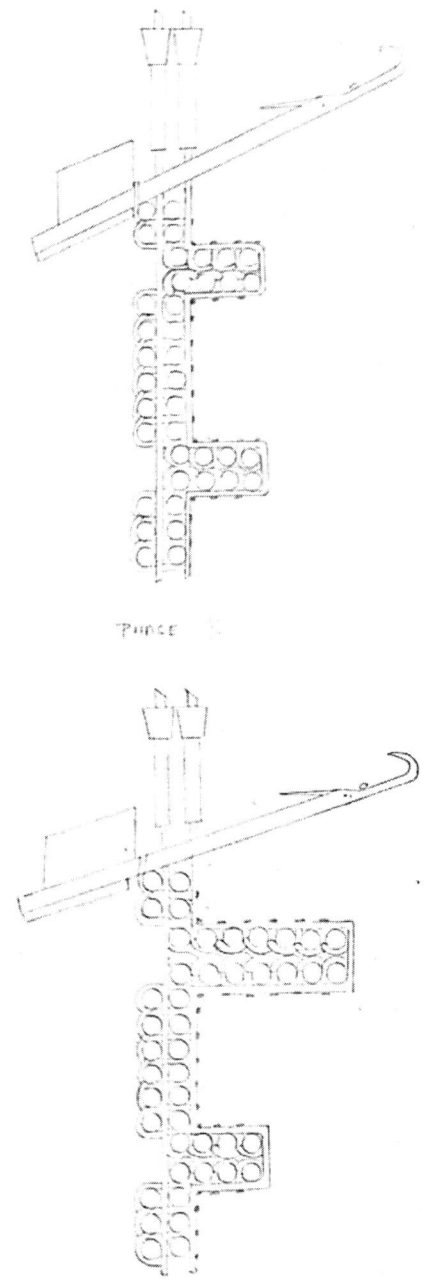

Figure 1 – New method of knitting of successive construction of perpendicular connected multiple layers(each comprising of 4 reinforcement layers [24]

Based on such new knitting method, two consecutive multi-layer reinforced knit structures, that have been produced with the same yarns and consist of individually four reinforcement layers, could be connected perpendicularly and the intermediate layer would hang independently (Figure 1). Based on this innovative knitting method, two consecutive multi-layer reinforced knit structures, which are produced using the same yarns and comprise individually four reinforcement layers, could be connected perpendicularly and the intermediate layer would hang independently.

Multi-layered 3D spacer fabrics with reinforcement layers integrating up to four reinforcement layers were developed based on the above-mentioned innovative knitting concept. The schematic manufacturing techniques of these spacer fabrics are documented in Figures 2 and 3.

This advanced flat-knitting technique for 3D spacer fabrics with four reinforcement layers also permits the manufacture in various curvilinear shapes in the warp direction.

The curvilinear shapes can be achieved by knitting the variable lengths of plane layers, while the connecting layers were knitted simultaneously on both needle beds.

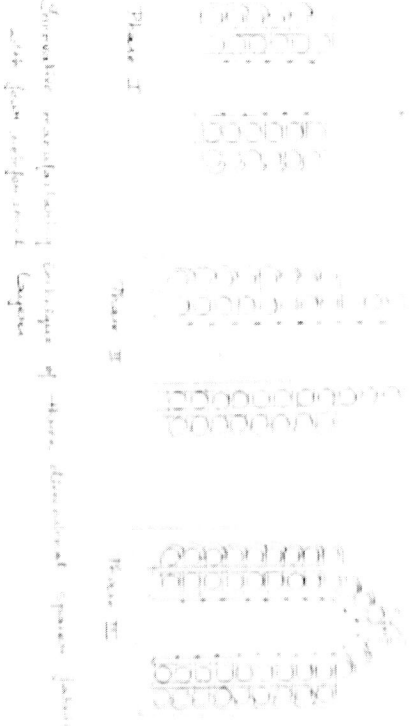

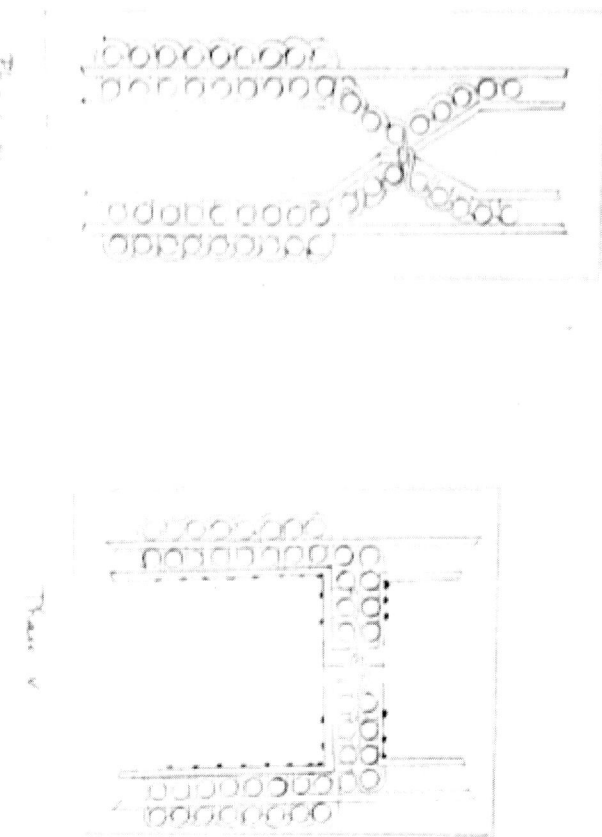

Figure 2 – Innovative production method of 3D spacer fabric having 4 reinforcement layers [24]

13.3   The findings

The flat-knitted new type of 3D spacer fabrics made from GF-PP filament hybrid yarns and having four reinforcement layers have been constructed. The 3D curved spacer fabrics of arc angle from 0° to 360° have been effectively designed, which establishes that the innovative manufacturing concepts are very flexible in shaping the spacer fabrics in the warp direction. The distance between the surface layers of all spacer fabrics have been designed to 30 mm. But in the case of spacer fabrics having 0° curvature the length between two neighboring connecting layers is 45mm. On the other hand in the case of curvature angles of 90° and 360°, it is dissimilar for both surface layers of the spacer fabrics because of inclined joining by the connection layers between the surface layers. The specific weight of fabric specific for surface as well as connecting layers is measured to about 0.62 kg/m^2.

Separate studies have been carried out relating to the mechanical properties of the 2D surface and connecting fabrics (also in the formation of thermoplastic composites) of these multi-layer reinforced 3D spacer fabrics and the results have been obtained [6,7]. Such findings clearly interpret the high performances of such multi-layered 3D spacer fabrics as textile preforms in composite applications.

The integrated 2D knit structures of the functional yarns and the multi-layered reinforced 3D spacer fabrics have been constructed. The structural condition of the end products can be supervised by means of functional or conductive materials that serve as sensor networks. Such innovative technology can result in economical production cost by minimization of the additional intermediate processes.

A considerable extension of the structural diversity of the knitting method is provided by the developed solution for complex-shaped multilayered innovative 3D spacer fabrics. As a result, it opens up the large-scale applications in the area of lightweight construction and energy technology. It opens up the opportunity for enhancement and design of the properties of knitted semi-finished products and the development of entirely new solutions. The other potential end uses comprise of energy-absorbing structures in protective textiles and crash-relevant components in automotive and mechanical engineering [24]. It can also be used in production of vehicle for the economical sandwich panel components (e.g. roof, door, deck, wing, under body construction) with short cycle times (about one minute in thermoplastic consolidation process). Also, the empty channels can be used as fuel or hydrogen tanks or their embodiment with sound-damping material. Considerable decrease in weight is the major merit of the structure. As a result, it appears to be highly prospective in the replacement of conventional metal-based structures. The resistance to pressure waves enabled by the energy absorption can be of more utility in buildings and military vehicles for blast protection.

Moreover, the innovative 3D spacer fabrics with high-performance fibers exhibit very good potential for application in civil engineering (lightweight building and bridges) and in architectural design, particularly as complex-shaped fiber reinforced lightweight building panel systems. The cavity chambers would be suitable as media, for instance, for maximum airflow, water passages, drainage systems, and power supplies.

On the other hand, the spacer fabric structures are supposed to exhibit better impact properties when the separation channels are equipped with the energy-absorbing materials. The ability to use a broad range of materials, full flexibilities in structural diversifications, and different reinforcing concepts, along with cost-effective single stage production render the latest flat-knitting technologies developed herein very prospective in the production of innovative 3D spacer fabrics for high-tech end uses, particularly in structures that are low in weight.

13.4   References

[1]   Savci S, Curiskis JI and Pailthorpe M. Knittability of glass fiber weft-knitted preforms for composites. Textil Res J 2001; 71: 15–21.

[2]   Vuure AWV, Ko FK and Beevers C. Net-shape knitting for complex composite preforms. Textil Res J 2003; 73: 1–10.

[3]   Abounaim M, Hoffmann G, Diestel O and Cherif C. Development of flat knitted spacer fabrics for composites using hybrid yarns and investigation of 2D mechanical properties. Textil Res J 2009; 79: 596–610.

[4]   Abounaim M, Hoffmann G, Diestel O and Cherif C. 3D spacer fabric as sandwich structure by flat knitting for composites using hybrid yarn. In: Proceedings of the Autex Conference, Izmir, Turkey, 2009, pp. 675-681.

[5]   Abounaim M, Hoffmann G, Diestel O and Cherif C. Flat knitted "spacer fabrics" with hybrid yarns for composite materials. Melliand Textileberichte 2008; 3–4: E30–31.

[6]   Abounaim M, Diestel O, Hoffmann G and Cherif C. High performance thermoplastic composite from flat knitted multi-layer textile preform using hybrid yarn. Comp Sci Technol 2011; 71: 511–519.

[7]   Abounaim M, Hoffmann G, Diestel O and Cherif C. Thermoplastic composite from innovative flat knitted 3D multi-layer spacer fabric using hybrid yarn and the study of 2D mechanical properties. Comp Sci Technol 2010; 70: 363–370.

[8]   Abounaim M, Hoffmann G, Diestel O and Cherif C. Thermoplastic composite from curvilinear 3D multi-layer spacer fabric. J Rein Plast Comp 2010; 29: 3554–3565.

[9]   Abounaim M. Process development for the manufacturing of flat knitted innovative 3D spacer fabrics for high performance composite applications. PhD thesis, Department of Mechanical Engineering, Technische Universita¨ t Dresden, 2011.

[10]   Hong H, Araujo M and Fangueiro R. 3D technical fabrics. Knitting Int 1996; 1232: 55–57.

[11]   Araujo MD, Hong H, Fangueiro R, Ciobanu O and Ciobanu L. Developments in weft-knitting technical textiles. In: Proceedings of the 1st Autex Conference(TECHNITEX 2001), Portugal, vol. 1, 2001, pp. 253–262.

[12]   Abounaim MD. Modelling of technical bindings and manufacturing of flat Knitted and woven "spacer fabrics" with hybrid (GF/PP) yarn as sandwich structure. Master's thesis No.-1310, Department of Mechanical Engineering, Technische Universita¨ t Dresden, Germany, 2006.

[13]   Ciobanu L. SANDTEX – developments on knitted sandwich fabrics with complex shapes. In: Proceedings of the 1st Autex Conference (TECNITEX 2001), Portugal, vol. 1, 2001, pp. 490–496.

[14] U¨ nal A, Hoffmann G and Cherif C. Development of weft knitted spacer fabrics for composite materials. Melliand Textileberichte 2006; 4: E49–50.

[15] Torun AR, Paul C, Hanusch J, Diestel O, Hoffmann G and Cherif C. Reinforced weft knitted preforms and spacer fabrics as well as woven spacer fabrics made of commingled hybrid yarns for RP. In: Proceedings of the Techtextil Symposium, Frankfurt, Germany, 12–14 June 2007.

[16] Cherif C, Ro¨ del H, Hoffmannn G, Diestel O, Herzberg C, Paul C, et al. Textile Verarbeitungstechnologien fu¨ r hybridgarnbasierte komplexe Preformstrukturen (Textile manufacturing technologies for hybrid based complex preform structures). Kunststofftech (J Plast Tech) 2009; 2: 103–129.

[17] Collaborative Research Centre. SFB 639, 'Textile-reinforced Composite Components for Function-integrating Multi-material Design in Complex Light weight Applications', Technische Universita¨ t Dresden, Germany, http://www.tu-dresden.de/mw/ilk/sfb639/sfb_en.html   (accessed   18 February 2011).

[18] Badawi SSAM. Development of the weaving machine and 3D woven spacer fabric structures for lightweight composites materials. PhD thesis, Department of Mechanical Engineering, Technische Universitat Dresden, 2007.

[19] Alagirusamy R and Ogale V. Commingled and air jet textured hybrid yarns for thermoplastic composites. J Ind. Textil 2004; 33: 223–243.

[20] Fujita A, Maekawa Z and Hamada H. Mechanical behavior and fracture mechanism of thermoplastic composites with commingling yarn. J Rein Plast Comp 1993; 12: 156–172.

[21] Sun BZ, Hu H and Gu BH. Responses of 3D biaxial spacer weft-knitted composite circular plate under impact loading (part i: unit-cell and elasto-plastic constitutive model). J Textil Inst 2010; 101: 28–34.

[22] Li JJ, Sun BZ, Hu H and Gu BH. Responses of 3D biaxial spacer weft-knitted composite circular plate under impact loading (part ii: impact tests and FEM calculation). J Textil Inst 2010; 101: 35–45.

[23] Zhang MX, Sun BZ, Hu H and Gu BH. Dynamic behavior of 3D biaxial spacer weft-knitted composite t-beam under transverse impact. Mech Adv Mater Struct 2009; 16: 356–370.

[24] Abounaim MD, and Chokri C, Flat-knitted innovative three-dimensional spacer fabrics: a competitive solution for lightweight composite applications, Textile research journal, 82(3)288.

Chapter 14

Warp knitted spacer fabrics in shoe insole applications

Summary

The warp knit spacer fabrics made of polyester filament have been investigated for permeability and conductivity properties. When the spacer fabric is used as shoe insole, it creases a comfortable fabric which normalizes the heat transfer while on foot or running. Such double-sided warp-knit spacer fabrics have been produced by varying the different thicknesses and face structures. In one set the spacer fabrics have been compared by maintaining the hexagonal net structure in face surface layer and the thicknesses of the fabrics have been changed to three different levels. In another set, the thickness of the spacer fabric has been kept constant and the face surface structure has been knit into locknit, rhombic mesh, and hexagonal net. The findings have shown that the air and water vapour permeability of the fabrics are influenced by the thickness and porosity. In the case of conductivity and permeability of spacer fabrics, the fabric porosity is considered as crucial. The vertical gap in the spacer fabric is determined face and back layers that comprise the two outer surface layers. The extent of porosity is determined by the contribution of middle layer in such vertical gaps. Good thermal conductivity results from the low vertical gap which indicates poor void space, and does not hence does not permit the air and moisture to penetrate to it. When the void space between the two fabric layers increases, the air gets trapped and the flow of air and moisture through the fabrics gets restricted by the middle layer. At a certain level of spacer fabric thickness optimum level of thermal conductivity is achieved with good air and water vapor permeability. The horizontal pore size of the spacer fabrics is determined by the surface structure of the face fabrics, which can be more open or closed structures. When compared with the other two structures, the open mesh hexagonal net structure exhibits good porosity. Good air and water vapour permeability are due to the open structure with long loop and moderate moving of number of filament. The properties of fabric considerably influence the the thickness and face layer structure as confirmed by ANOVA technique. The Tukey's HSD test also proved the significant difference among all the groups. From the above considerations, the fabric having open structure with around 3-mm thickness is found to be suitable for shoe insole with high air and moisture permeability that can balance the thermal conductivity.

14.1   Introduction

In the area of sports textiles comfort is considered to be important aspect. Many researchers had turned their attention on the foot wear. Whether on foot or running shoes evacuate the warmth created in the foot. Conduction, convection, and radiation are the important means of heat transfer. The heat convection takes place by means of the aeration of the gap and also by evaporation, which constitute the basic mode of heat transfer between the foot and the ambient air [1]. The insole is used to decrease the force transmission and customizing shoe to protect the foot and cause decrease in the occurrence or recurrence of heat [2]. In order to protect against discomfort the cushioned or shock-absorbing insoles have been recommended as a material to diminish the impact forces related to running and decrease plantar pressures [3,4]. Studies have shown that the cushioned insoles mitigate the risk of stress fractures and overuse injuries [5,6]. With the introduction of the insole materials, the pressure at the painful area decreases below 254.97 kN/m2 [7]. Prior investigations have shown that insoles made of only polyurethane foam and neoprene go through a major deterioration in shock-absorbing capacity after a few weeks of daily walking. The insole has been made with 3-mm thickness at the forefoot as well as at the center of the heel [8,9]. Despite the shock-absorbing capacity of the insole being considered as the main criteria, the wearing comfort also plays an important part. For obtaining this breathable wearing comfort, the foams may not be appropriate and the other option would be a textile material. The construction of textile materials is highly complicated and covers a broad range. Whereas, the structure and material composition are considered to be the major parameters for determination of the permeability and comfort properties. In the case of cushioning applications, warp-knitted spacer fabrics are an ideal group of energy absorbers. Their energy absorption ability can easily be customized for satisfying proper needs by merely altering their structural factors [10,11]. The basic construction of 3D spacer fabrics is formed of two surface textile layers held by spacer threads in a defined spacing. Such structure enables spaces which easily permit heat and moisture to be transferred through the fabric [12,13]. The quantity of yarn contribution in a distinct unit area is based on the fabric structure and thickness, which determines the loop length, stitch density, and areal density. The production of spacer fabrics normally involves the polyester multifilament yarns on its two outer surfaces and monofilament yarns at the middle [14,15]. Such comfort property of the spacer fabric is based on the fabric thickness and structure. The thickness and weight of the fabric play a considerable role, as it determines the distance through which moisture vapor and heat passes from one side of the fabric to the other [16]. Owing to metabolic processes of the human body heat is generated continuously and gets transferred through the clothing. The thermophysiological comfort characteristics relate to mainly the fabric transmission performance and maintan the heat balance between the body and the surroundings. The efficiency of heat dissipation is related to fabric permeability and conductivity properties. Usually, the spacer fabrics show good thermal conductivity, which are determined by the weight, structure, and thickness of the fabric. Thermal resistance corresponds to the fabric thickness and often has an inverse relationship with the thermal conductivity [17]. The number of pores in the fabric has a direct bearing on the permeability properties of fabrics. The porosity of the fabric determines the quantity of air entrapped within the fabric construction and the air passing through the fabric [18,19]. The comfort property is dependent on air and water vapor permeability which are also the most significant properties of textile materials. The conditions of measurement do not affect the surface porosity on water vapor permeability and air permeability [20,21]. In general, the mechanical behaviors like tensile, tear, and peeling are the other attributes of spacer fabrics and it can be utilized in an innumerable applications varying from cushioning to automotive applications, active wear to extreme sporting apparel, intimate wear, medical, and wide range of industrial applications [22,23]. Attention has been focused on the study the influences of spacer fabrics for the application of shoe insole with supernatural properties of permeability and conductivity. Warp-knitted polyester spacer fabrics have been produced so as to study the fabric porosity, air permeability, and water vapor permeability. One set of fabrics has been designed with hexagonal net structure on face surface layer having three fabric thicknesses. Another set of spacer fabrics have been produced with thickness of 3 mm with varying three face surface layer of close and open structure as the face layer is having contact to the skin.

14.2 Technical details

The warp knit spacer fabrics have been knitted from polyester multifilament and monofilament yarns of various linear densities. The middle layer of the spacer fabric comprised of monofilament yarn, and a certain gap has been maintained between the two outer surface layers. The spacer fabrics have been knitted on a raschel type double needle bed warp knitting machine with particular machine guage and guide bars. The fabrics have been knitted with varying thickness. Two sets of spacer fabrics have been produced. The first set comprised of 3 levels of thicknesses using the hexagonal net structure on face surface and plain structure at bottom surface layer. The second set has been produced with a certain thickness, and the structure of the face surface layer is locknit, rhombic mesh and hexagonal net and plain in bottom surface.

Fabric thickness guage has been used to measure the thickness of the fabrics as per prescribed test methods. Weighing balance has been used to measure the areal density as per prescribed method. Fabric weight has been calculated.

The air permeability, moisture permeability, and thermal conductivity of a fabric is influenced by porosity. It was estimated by using a formula. Prescribed test method has been used to measure the air permeability of the test fabrics. Lee'disk instrument has been used to calaulate the thermal conductivity. The thermal resistance and relative water vapor permeability have been measured on permetest instrument working on similar skin model principle based on prescribed BS and ISO standards. The relative water vapor permeability of the fabrics has been calculated by means of an equation.

14.3   Influence of porosity

The porosity of a spacer fabric is a crucial aspect that determines its permeability, moisture, and thermal comfort. The factors such as loop length, stitch density, and thickness of the fabric have a major influence on its porosity [24]. The porosity of the spacer fabric is based on its structure and thickness, with a constant linear density of polyester filament in the three layers. The middle spacer fabric layer determines the space between the two outer surface layers. Despite having the similar surface structure on the face, of the three test fabrics(having different thicknesses, the minimum and maximum vertical gap of spacer fabric leads to a low porosity compared with the fabric having intermediate thickness. The fabric having lowest thickness does not contain enough space between the two surface layers and exhibits poor porosity. The fabric having highest thickness has more space which results in more contribution of middle layer and increase in fabric density leads to low porosity. The stitch density and the structure of the fabric determine the horizontal pore size. With a locknit structure the filaments are inter looped very closely in comparison with than the other two open structures. Long loop is produced by the open structure on the surface, and the relative porosity increases by removal of few filaments. It is made evident in Figure 1 and there is a steady improvement in the value of porosity between fabrics with locknit structure and hexagonal structure. The horizontal increase in porosity considerably increases in air and water vapor permeability.

14.4   Air permeability of fabric

Air permeability is a measure of the air flow rate passing vertically through a defined area of fabric under a suitably adjusted air pressure differential between the two surfaces of a material. It directly relates to porosity. A fabric with a very high porosity can be considered to be permeable [18,25]. From figure 1 it can be observed that the fabric having intermediate thickness and fabric having rhombic mesh structure exhibit more open structure compared to the others and with the increment of porosity, the air permeability values have two structures. When the fabric with rhombic mesh structure produces open structure on its surface, it creates more spaces and it permits sufficient air to pass through the fabric.

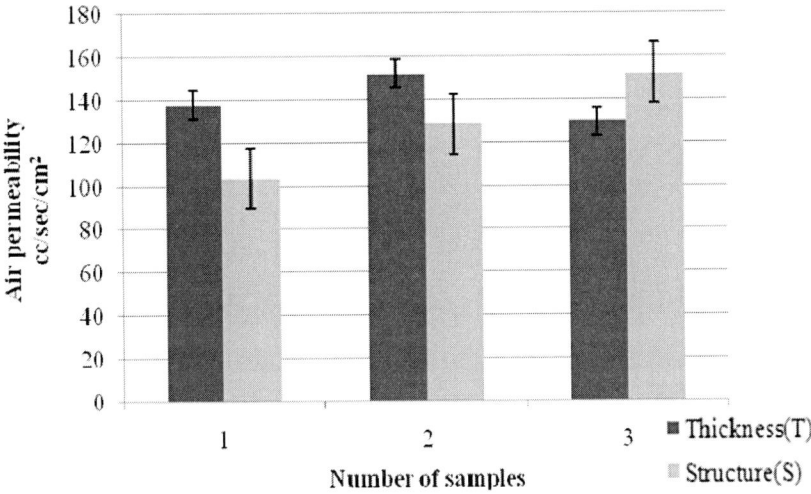

Figure 1 – Air permeability of test fabrics [29]

14.5   Water vapor permeability of fabric

One of the crucial aspects that decide the velocity of water vapor transmission through a textile material is the water vapor permeability. It is a critical factor in evaluation of comfort properties of a fabric, since it pertains to the ability of transporting perspiration. Overall, the polyester superior water vapor permeability which is because of low tendency of retaining moisture within the filament [26]. Figure 2 depicts the water vapor permeability of various kinds of warp-knitted polyester spacer fabrics. In the case of highly open knit fabric structure the moisture vapor transmission is predominately dictated by fabric variables which decide the thickness and permeability. The fabric thickness is a crucial aspect and it establishes the distance through which moisture vapor pass through both sides of the fabric. The less space in fabric having lower thickness and more space in fabric of higher thickness hold the moisture within it, and results in low permeability. The structural aspect of the fabric also has an effect on the moisture vapor performance. In this case, the open structure fabric of higher thickness has more water vapor diffusivity in between the surfaces and layers. When there are more number of pores in the fabric structure, it leads to high porosity, and thus good water vapor permeability. Such an amount of permeability is termed as the property of a porous material and characterizes the passage of liquid that is forced to flow through the fabric under an applied force. The fabric having lower thickness and compact structure factor considerably affects the relative water vapor permeability. So the unfasten structure forms a transfer system that draws moisture from the skin to the outer layer of the fabric.

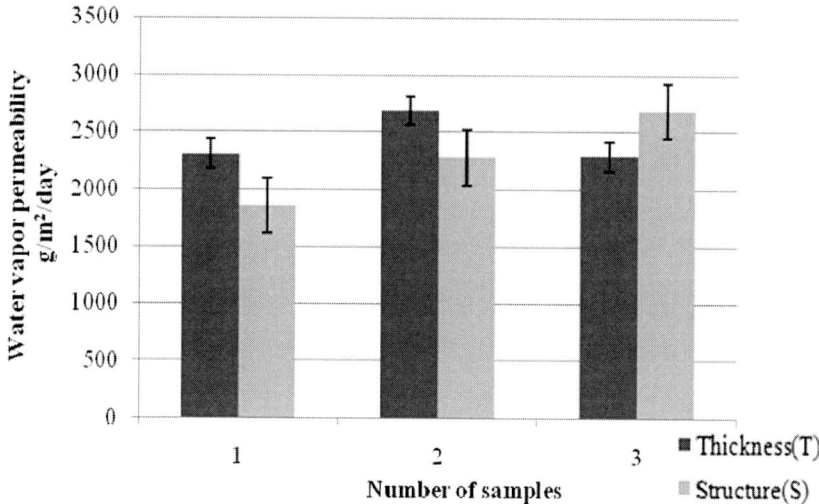

Figure 2 – Water vapour permeability of test fabrics [29]

14.6 Thermal properties of fabric

Thermal properties have been described as the amount of heat transferred through the fabric thickness in a measured surface area. The thickness of the fabric and its structure has a strong effect on the thermal conductivity and resistance. The least thermal conductivity is seen in the fabric having long float open loop structure with higher thickness. The thermal resistance of the fabric exhibits a very good response for thermal conductivity and thickness [27,28].

The influence of thickness and structure on the thermal conductivity of the fabric is depicted in figures 3 and 4. The comfort property is based on the fabric thickness, and thermal conductivity has been established as a considerable factor resulting in the thermal insulation of textiles. Such findings give details in the manner that comparatively greater fabric thickness of a spacer fabric entraps more air within the middle layer and hence cause greater thermal resistance with lower thermal conductivity.

Despite the fabric having maximum thickness has high middle layer density among the three test fabrics, the greater fabric thickness contributes more on thermal properties. When taking into account fabric design, the findings can be described by the structure of the outer surface layers. The fabrics with open skin hexagonal net structure exhibit comparatively lower thermal conductivity in comparison with the closed skin locknit structure. The amount of entrapped air within the hexagonal net structure is high and it limits the free transmission of heat leading to lower thermal conductivity and greater thermal resistance.

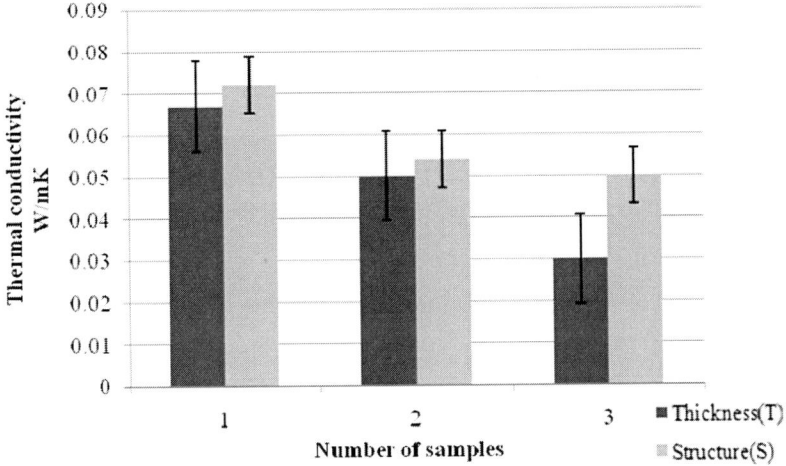

Figure 3 – Thermal conductivity of test fabrics [29]

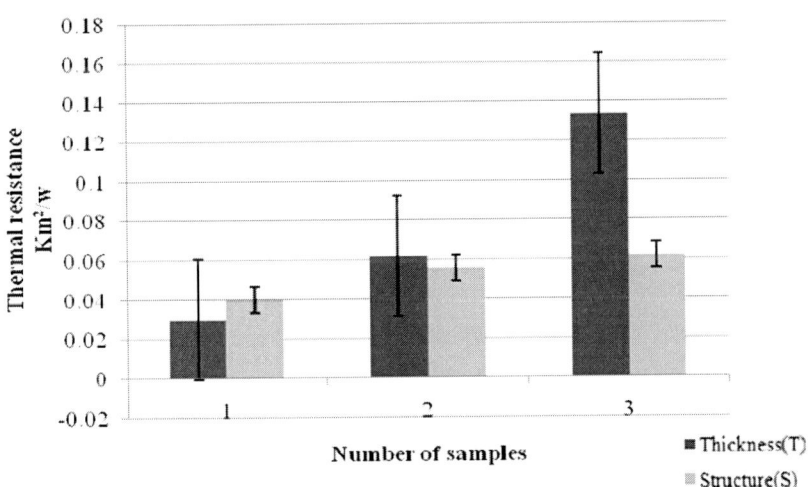

Figure 4 – Thermal resistance of test fabrics [29]

14.7   Statiscal analysis

ANOVA confirms the aforesaid findings, and shows considerable effect of the structure and thickness on fabric properties. One-way ANOVA has been studied and the chosen value of significance for all statistical tests in the study is 0.05 level. The degree of freedom is 2, 12, and the Fcritical is 3.89. The results of the ANOVA have been determined, and evaluation has been done on the influence of groups of thickness and structure of test spacer fabrics with regard to porosity, thermal properties, air, and water vapor permeability. The value of Fcritical<Factual proves that the changes in the thickness and surface layer structure of warp-knitted spacer fabric results is highly significant on the above-mentioned fabric basic comfort properties. Even the minor changes in the fabric thickness leads to in significant impact on the fabric properties. It has also been established that the change in surface layer structure reflects very high influence on the fabric properties in comparison with thickness. Turkeys honest significance difference (HSD) test has also been studied so as to confirm the significance of ANOVA. Tukey's HSD test is used to determine which groups in the test fabric differ, and the test shows that there is a significant difference among all the groups. The Q(3,12) values are 3.77 and Q critical<Q actual for all the paired mean differences. Such study establishes that all the warp-knitted spacer test fabrics show considerable difference in porosity, air, and water vapor permeability.

14.8   References

[1]   Rebay M, Arfaoui A and Taiar R. Thermo-mechanical characterization of the interaction foot-athletic shoe during the exercise. 5th European thermal-sciences conference, The Netherlands, 2008.

[2]   Brodsky JW, Pollo FE, Cheleuitte D, et al. Physical properties, durability and energydissipation function of dual-density orthotic materials used in insoles for diabetic patients. Foot Ankle Int 2007; 28: 880–889.

[3]   Forner A, Garcia AC and Alcantara E. Properties of shoe insert materials related to shock wave transmission during gait. Foot Ankle Int 1995; 16: 778–786.

[4]   Nigg BM, Herzog W and Read LJ. Effect of viscoelastic shoe insoles on vertical impact forces in heel-toe running. Am J Sports Med 1988; 16: 70–76.

[5]   Schwellnus MP, Jordaan G and Noakes TD. Prevention of common overuse injuries by the use of shock absorbing insoles: a prospective study. Am J Sports Med 1990; 18: 636–641.

[6]   Mundermann A, Stefanyshyn DJ and Nigg BM. Relationship between footwear comfort of shoe inserts and anthropometric and sensory factors. Med Sci Sports Exerc 2001; 33: 1939–1945.

[7]   Leber C and Evanski PM. A comparison of shoe insole materials in plantar pressure relief. Prosthet Orthot Int 1986; 10: 135–138.

[8]   Pratt DJ. Medium term comparison of shock attenuating insoles using a spectral analysis technique. J Biomed Eng 1988; 10: 426.

[9]   O'Leary K, Vorpahl KA and Heiderscheit B. Effect of cushioned insoles on impact forces during running. J Am Podiat Med Assoc 2008; 98: 36–41.

[10]   Liu Y, Hu H, Zhao L, et al. Compression behavior of warp-knitted spacer fabrics for cushioning applications. Tex Res J 2011; 82: 11–20.

[11]   Liu Y, Hu H, Long H, et al. Impact compressive behavior of warp-knitted spacer fabrics for protective applications. Tex Res J 2012; 82: 773–788.

[12]   Bagherzadeh R, Montazer M, Latifi M, et al. Evaluation of comfort properties of polyester knitted spacer fabrics finished with water repellent and antimicrobial agents. Fibers Polym 2007; 8: 386–392.

[13]   Davies A and Williams J. The use of spacer fabrics for absorbent medical applications. J Fiber Bioeng Inform 2009; 1: 321–330.

[14]   Ye X, Hu H and Feng X. An experimental investigation on the properties of the spacer knitted fabrics for pressure reduction. RJTA 2005; 9: 52–57.

[15]   Ye X, Hu H and Feng X. Development of warp knitted fabrics for cushion application. J Ind Tex 2008; 37: 213–223.

[16]   Crina B, Blaga M, Luminita V, et al. Comfort properties of functional weft knitted spacer fabrics. Tekstil ve Konfeksiyon 2013; 23: 220–227.

[17]   Prahsarn C, Barker RL and Gupta BS. Moisture vapor transport behavior of polyester knit fabrics. Tex Res J 2005; 75: 346–351.

[18]   Onal L and Yildirim M. Comfort properties of functional three-dimensional knitted spacer fabrics for home-textile applications. Tex Res J 2012; 82: 1–14.

[19]   Tugrul Ogulata R and Mavruz S. Investigation of porosity and air permeability values of plain knitted fabrics. Fibers Tex East Eur 2010; 18: 71–75.

[20]   Wilbik-Halgas B, Danych R, Wie_cek B, et al. Air and water vapour permeability in double-layered knitted fabrics with different raw materials. Fibres Tex East Eur 2006; 14: 77–80.

[21]   Palani Rajan T, Souza LD, Ramakrishnan D, et al. Influence of porosity on water vapor permeability behavior of warp knitted polyester spacer fabrics. J Ind Tex 2014; 1–17.

[22]   Ming-Xing Z and Hong HU. The mechanical behavior of 3D bi-axial spacer weft knitting fabrics. J Xi' Poly Uni 2009; 23: 285–291.

[23]   Delkumburewatte GB and Dias T. Porosity and capillarity of weft knitted spacer structures. Fibers Polym 2009; 10: 226–230.

[24]   Guo X, Long H and Zhao L. Investigation on the impact and compression-after-impact properties of warp-knitted spacer fabrics. Tex Res J 2013; 83: 904–916.

[25] Yip J and Ng S-P. Study of three-dimensional spacer fabrics: physical and mechanical properties. J Mater Process Tech 2008; 206: 359–364.

[26] Shabridharan and Das A. Study of heat and moisture vapor transmission characteristics through multilayered fabric ensembles. Fibers Polym 2012; 13: 522–528.

[27] Onofrei E, Rocha AM and Catarino A. The influence of knitted fabrics' structure on the thermal and moisture management properties. J Eng Fibers Fabrics 2011; 6: 10–22.

[28] Chidambaram P, Govind R and Venkataraman KC. The effect of loop length and yarn linear density on the thermal properties of bamboo knitted fabric. AUTEX Res J 2011; 11: 102–105.

[29] Palani rajan T, and Ramakrishnan G, 2016, Comfort properties of functional warp-knitted Polyester spacer fabrics for shoe insole applications, Journal of industrial textiles, 45(6)1239.

Chapter 15

Weft knit spacer fabrics in medical application

Summary

The important needs of wound dressings in the case of pressure ulcer have been taken into account. In addition to good absorption property, breathability, thermal regulation and cushioning properties are also required to satisfy the wound dressing qualities for pressure ulcer.. Hence, the air permeability, thermal conductivity, water vapor permeability, absorbency and compression of weft knitted spacer fabrics have been studied and compared with existing wound dressings so as to assess whether the former can be used as a substitute for the latter. The results have shown that because of their structure the air as well as water vapor permeability of spacer fabric proves to be far better than those of other wound dressings. Though the thermal conductivity of spacer fabrics do not compare well with all dressings, they are still comparable with some of the dressings considered. It also holds valid in the case of the absorbency performance, and their absorbency property is comparable with certain dressings considered. The findings also make it evident that compression resistance as well as compression resilience of some weft knitted spacer fabrics are better than that of wound dressings.

15.1   Introduction

Pressure ulcer results from the damage of skin or underlying tissue owing to continued pressure over long duration or pressure integrated with sheer and friction over a bony prominence. The heel is considered as one of the most common sites for pressure ulcer development [1-5]. Despite wound dressings being used over years, management of wound is becoming highly complex [6-7]. The objective of modern wound dressings is to improve the healing of wounds [8-9]. Also, the prevention and treatment of pressure sores require considerable time and care. It greatly affects the lives of patients and their caretakers as well as the hospital services and costs of government since long term hospitalizations are necessary and prove expensive. Earlier work has established that the health costs of pressure ulcers are undoubtedly high [3,10,11]. Pressure ulcers do not easily heal and wound dressings which offer both good absorption as well as cushioning effect are not common. Spacer fabrics are basically 3D structures consisting of upper and lower fabric layers having filament yarn in between to connect them together by tuck loop stitching [1,12,13]. Owing to their versatile physical properties, weft knitted spacer fabric has attracted good deal of attention [14]. As revealed from earlier work, physical properties of spacer fabrics can be easily adjusted through use of different kinds of spacer yarn, fabric density, thickness and fabric structure [15,16]. Because of their versatile physical, mechanical and thermal properties, it has been found to have a wide area of applicability, including in medical products [17,18]. Their excellent ventilation and cushioning properties are also important for pressure ulcer prevention and the healing process [19,20]. The essential physical properties of the absorbent layer of wound dressings, including air permeability, thermal conductivity, water permeability, absorbency and compression, have been studied in 3-dimensional weft knitted spacer fabrics, and then assessed and compared with those of wound dressings obtained commercially. The objective has been to develop a better understanding of the specific physical properties of weft knitted spacer fabrics designed for use as the modern wound dressings.

15.2   Technical details

Fifteen types of weft knitted spacer fabrics that are commercially available have been considered for the investigation. Out of these eleven test fabrics have similar structure while remaining have different structures. The weft knit spacer fabrics had 100% polyester in one fabric and varying blend compositions of polyester and elastane in others(Figures 1 and 2). Parameters like fabric structure, angle of spacer yarn, areal density and bulk density have been determined. Also, seven different types of wound dressings that are particularly used for burns, ulcers and surgical wounds with cushioning effect have been chosen for the sake of comparison. The technical details of the wound dressings are given in Table 1 below

Table 1 – Technical details of wound dressings selected [1]

Dressing	Type	Material description
Type 1	Non adhesive	Absorbent hydrocellular pad sandwiched between a perforated non-adherent wound contact layer and a water proof outer film.
Type 2	Non adhesive	Polyurethane foam dressing which offers a good healing environment for moist wounds and effective exudate management for fragile skin particularly pressure ulcers.
Type 3	Self adherent	A highly absorbent material embedded in a self adherent polyurethane matrix for the management of extruding wounds surrounded by intact skin.
Type 4	Non adhesive	Consists of a hydrophilic polyurethane membrane matrix.

Type 5	Adhesive	Consists of hydrocolloids within the dressing while the adhesive layer contains polymers which will form cohesive gel when it comes into contact with wound exudates.
Type 6	Non adhesive	Can be used for low to moderately exuding leg and pressure ulcers
Type 7	Adhesive hydrocolloid	Used for pressure ulcers and other moderately exuding wounds. It is composed of carboxymethylcellulose particles in an elastomer network. When it comes into contact with exudate, it will form a soft moist gel that can enhance the wound healing process.

The tests have been based on ASTM standards under standard conditions. The materials have been tested for air permeability, thermal conductivity, water vapor permeability, absorbency, and compression.

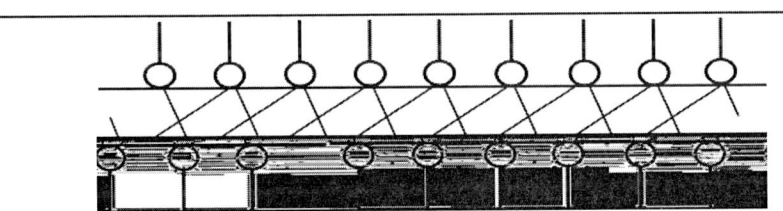

Figure 1 – Technical diagram of weft (100% polyester) to weft (86% polyester 14% elastane) [27]

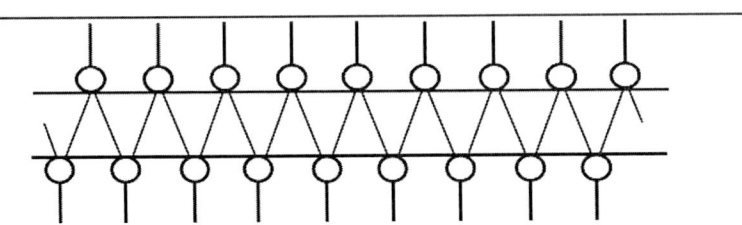

Figure 2 – Technical diagram of weft (93% polyester, 7% elastane) to weft (87% polyester, 13% elastane) [27]

15.3   Studies on air permeability

Figure 3 depicts the results of air resistance (R) of all the spacer fabrics and wound dressings. With the increase in the value of R the air permeability of the test fabrics decreases. The air permeability of fabrics is influenced by important parameters that include fabric density, thickness and tightness

[16]. Based on the results, no data were recorded for all Dressings even when the largest hole and the maximum range were used as their air resistance exceeded 500 kPa·s/m. The findings reveal that wound dressings exhibit very low air permeability. The type1 dressing is the thickest among all the test samples while other dressings of type 3, 5, 6 and 7 exhibit greater bulk density and even have an adhesive layer. Hence, air cannot pass through them easily leading to low air permeability.

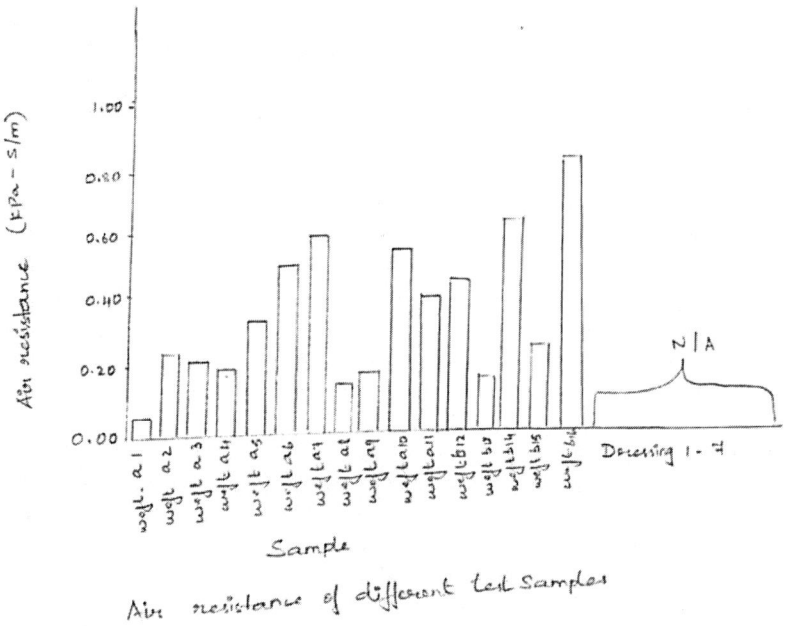

Figure   3Air resistance of different samples[27]

The weft knitted fabric having 100% polyester exhibits better air permeability in comparison with other weft knitted spacers since its thickness is relatively lesser compared to others. Since the weft knit spacer fabric having 87% polyester and 13% elastane exhibits the highest bulk density among all weft knitted spacer fabrics, its air permeability is the lowest. Even though the weft knitted spacer fabric has two different structures, only the interlaying structures of them are different while the face as well as the back fabrics is plain knit. Hence, the pores on their fabric are alike. Their variation in air permeability arises due to thickness, bulk density and the angle of spacer yarn. Air permeability is one of the main fundamental necessities of an ideal wound dressing that can offer a ventilated environment for wounds. Based on the discussion of findings above, the air permeability of the weft knitted spacer fabrics are found to possess higher air permeability compared with existing wound dressings.

15.4   Thermal conductivity

Figure 4 depicts the findings of thermal conductivity of all the test fabrics. A higher thermal conductivity value shows that the test fabric can conduct heat away from skin and body quicker [16, 21,22]. Earlier research showed that the most significant influencing factors of the thermal conductivity of the test fabrics are conductivity of individual fibers, fabric thickness, bulk density and type and angle of spacer yarn [16,23,24]. All the spacer fabrics are made of polyester, and hence, the differences in thermal conductivity arise because of their thickness, bulk density and angle of spacer yarn. The findings reveal that there is increase of the thermal conductivity of spacer fabrics with increasing fabric thickness, density and angle of spacer yarn since there are less spacer to trap air inside or improve the ventilation of heat. Weft knit spacer fabrics having 88% polyester and 12% elastane has greater thermal conductivity and followed by Weft knit spacer fabrics having 90% polyester and 10% elastane, and 89% polyester and 11% elastane, 86% polyester and 14% elastane respectively since they have greater thickness among weft knitted spacer fabrics. Depending on the results achieved, the thermal conductivities of the spacer fabrics are found to be same as each other, but lower than those of the dressings, particularly dressings conforming to types 3, 6 and 7 since the dressings have extremely high bulk density compared to others.

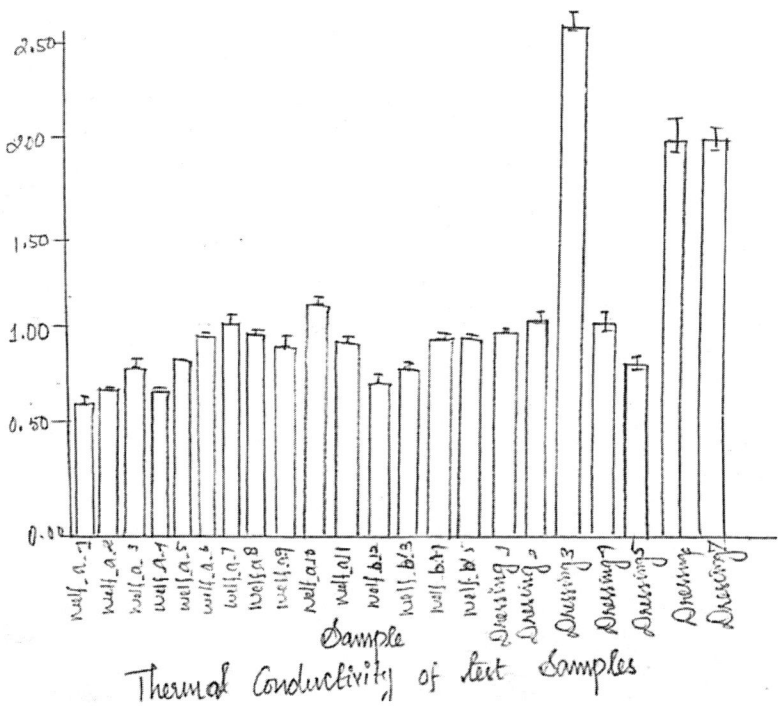

Figure 4 – Thermal conductivity of test samples[27]

15.5 Water vapor permeability

The water vapor permeability has been used for the assessment of the penetration of water vapor from skin to outside through the test sample, and figure 5 depicts the findings. Greater value of water vapor permeability shows more water vapor can pass through the sample to outside. Earlier investigation has shown that WVP performance is highly correlated with bulk density, thickness and fabric structure [21]. Depending on the findings, the WVP of spacer fabrics is always better compared with those wound dressing showing that more water vapor can penetrate through the spacer fabrics since the pores on the spacer fabrics enables the penetration of water vapor. Weft knitted spacer having second category structure exhibits better WVP compared with most of the spacers having first category structure. Also, as the bulk density and thickness increase, the WVP is reduced since more water will be trapped inside. Whereas, with the smallest bulk density among dressings, dressings types 1 and 2 have relatively high WVP that enables water vapor to penetrate through it. The other dressings have extremely small value of WVP, besides their greater bulk density, and having an adhesive layer will obstruct the penetration of water vapor.

WVP is considered as one of the fundamental requirement of wound dressings since it can maintain ventilation of wound by transmission of wound exudates and sweat from surrounding area. The findings reveal that the average WVP of the spacer fabrics is comparable or even better than that of the wound dressings.

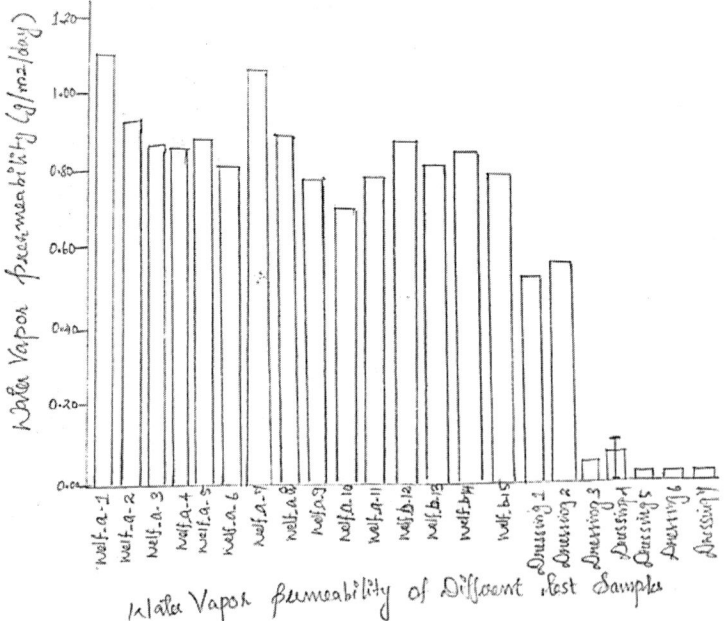

Figure 5 – Water vapour permeability of different test samples [27]

15.6   Absorbency

Figure 6 depicts the findings of absorbency test of water and 0.9% saline water wherein a greater value of absorbency implies that more water can be absorbed. The spacer fabrics as well as dressings exhibit poor absorbency ability in 0.9 saline solution rather than water as the sodium chloride particles in 0.9% saline water is larger than that of water particles. During the initial phase of absorption, small water particles will enter the sample because of the presence of the concentration gradients between solution and samples. But, as a hypertonic solution is formed, the water vapor particles cannot further enter the samples but start to move out from sample to the solution. Hence, the absorbency capacity of all test samples in 0.9% saline solution is lower than that of pure distilled water. Based on the findings, type 1 dressing that is made of a highly absorbent material (hydrocellular polyurethane) exhibits the highest absorption capacity among all samples since it can absorb the fluid directly into themselves rather than into the space between the fibers. Type 2 dressing is made of polyurethane foam as well.

But, its thickness is lesser than type 1 dressing, and there is fewer space inside it which makes its absorbency lower compared with type 1 dressing. Type 4 dressing has good absorption capacity since it is made of a thick layer of absorbent materials, hydrophilic polyurethane matrix. Even though dressing types 3, 5, 6 and 7 are also made of absorbent materials that are hydrocolloid and hydro-active polyurethane matrix, they exhibit poor absorbency. It is due to the fact that their surfaces have an adhesive layer that can greatly decrease its absorbency as the film prevents the dressing from absorbing wound extrudes. In the case of spacer fabrics, their absorbency is similar since they are made of polyester fibres that exhibit low water absorbency. The differences between them arise from their bulk density and thickness.

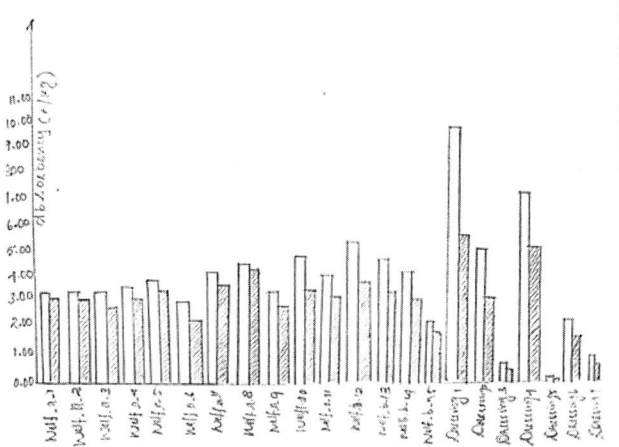

Figure 6 – Absorbency of spacers and wound dressings tested by using water and 0.9% saline solution [27]

A very important need of wound dressings is absorbency since various wounds can have various degree of exudates. Moreover, the number of times the wound dressing requires to be changed is influenced by the absorbency performance. Depending on the findings, the spacer fabrics have higher absorption ability compared to that of dressings having an adhesive layer (dressing types 3, 5, 6 and 7). Even though spacer fabrics used for the investigation are produced from polyester whose absorbency is lower compared to those wound dressings produced from highly absorbent materials, the their wicking property can enable to draw extrudes from wound through fabrics to outside rather than trapping inside the wound dressings so as to inhibit the possibility of bacteria growth.

15.7   Compression

Earlier literature has established that pressure higher than 32 mmHg (around 4.27kPa) is normally considered as the breakdown pressure that is higher th the capillary closing pressure in healthy individuals [25,26]. In addition, depending on the heel interface pressure measured before, the means of heel interface pressure of elderly when they are at 60 degrees and 90 degrees to the standard mattress are 50.15mmHg and 78.71mmHg respectively. Also, the maximum heel interface pressure of them on standard mattress at 90 degrees is 108.75mmHg [1]. Hence in the investigation, more emphasis has been given to the compression behavior of spacer fabrics at 4.27kPa (32mmHg), 6.69kPa (50.15mmHg), 10.47kPa (78.71mmHg) and 14.50kPa (108.75mmHg). Figures 7 & 8 depict the compression stress-strain and compression stress-thickness curves of all the spacer fabrics.

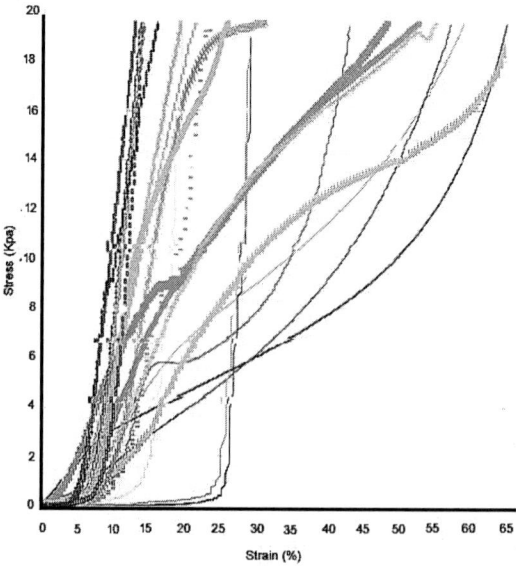

Figure 7 – Results of stress-strain curves of knitted spacer fabrics (left)[27]

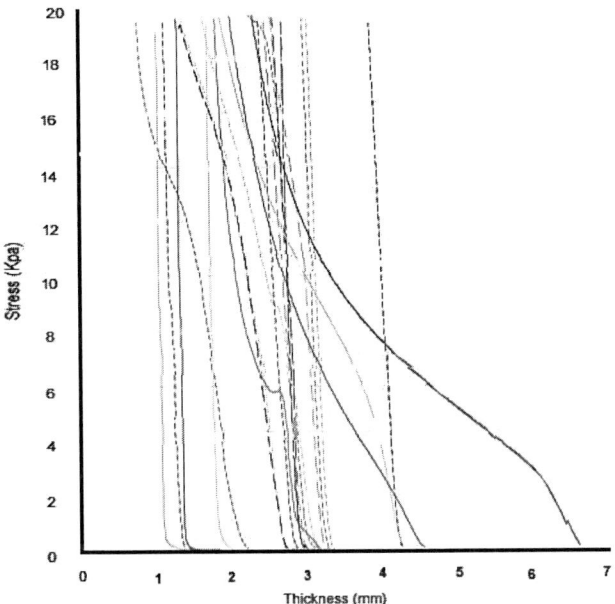

Figure 8 – Results of stress-thickness curves of knitted spacer fabrics (right) [27]

The findings reveal that as the angles of spacer yarns in the course-wise direction increase the compression resistance increases as established by earlier research [4]. On the other hand, the findings also show that the compression resistance is influenced by the percentage of polyester. The findings reveal that weft knit fabrics having 100% polyester monofilament weft, 94% polyester and 6% elastanc, 93% polyester 7% elastane(two different areal densities), respectively, with highest percentage of polyester content, exhibit lowest compression resistance. Similar trend is also noticed in other weft knitted spacer fabrics. Even though weft knit spacer fabric having 86% polyester and 14% elastane has the lowest percentage of polyester, its compression resistance is not the best one among weft knitted spacer fabrics. It is so since its spacer yarn angle in course-wise direction is not large. Hence, besides composition percentage, spacer yarn angle in course wise direction is also a critical factor. Even though weft knit spacer fabric with 89% polyester and 11% elastane do not have the lowest polyester percentage and their spacer yarn angles are not the largest, their compression resistance is good.

It arises from the relatively large fabric thickness in comparison with others that largely improve their compression resistance. It also holds valid in the case of weft knit spacer fabrics having 89% polyester and 11% elastane (having three different thicknesses), and which have the same polyester content (89%). The spacer yarn angle in the first two cases is not as large as the other (89% polyester and 11% elastane), but, their thickness is relatively larger. Four types of fabrics are having the same percentage of polyester (93%) but with different interlacing structure. One out of the four has the best compression resistance while the other three are similar. It arises due to the spacer yarn angle in course-wise direction of weft knit spacer fabric having 93% polyester and 7% elastane is the largest(higher areal density fabric) while that of other with similar polyester content(lower areal density) is the smallest. Hence, the compression resistance of Weft knit spacer fabric with 93% polyester content and lower areal density is the poorest.

It is obvious that weft knitted spacer fabrics always have better compression resistance in comparison with existing wound dressings at various pressure levels. Further, the resilience of such weft knitted spacer fabrics are good since the area between two curves of it is small.

It also holds valid for dressing types 3, 6 and 7. Whereas, dressing types 1, 2, 4 and 5 have large area between curves that show their compression resilience is poor. Dressing curves 1 and 2 show similar compressional curves since both of them are polyurethane foam having similar thickness and bulk density. Type 1 dressing exhibits the best compression resistance compared to others when pressure is under 2kPa. However, its compression strain increases rapidly when gradually increasing the pressure which implies its sponge layer cannot absorb much pressure. It also holds valid in the case of type 2 dressing that has similar compression resistance like weft knitted spacer fabrics when pressure is under 2kPa. Even though types 3 and 4 dressings are also made of Polyurethane matrix, the bulk density of type 3 dressing is 5.5 times greater than that of type 4 dressing that renders it stiffer to withstand pressure.

The change in fabric thickness under compression stress is depicted in figure 8. It can be seen that there is only a small thickness change in weft knitted spacer fabrics, which implies that weft knitted spacer is not easily compressed even under a pressure of 14.50kPa because of the presence of the spacer yarn between the fabrics. Whereas, type 1 dressing exhibits higher change in thickness than those of other types of dressings. There is reduction in its thickness around 3.5 mm under a force of 14.50kPa since type 1 dressing has a foam structure that is not stiff. It also holds valid in the case of dressing types 2 and 4. Whereas, the thickness reduction with type 5 dressing is not obvious as it is a very stiff material and cannot be easily compressed. For Dressing 5, there is a turning point at nearly 6kPa as its two different layers have different compression properties. Even though type 6 and 7 dressing are also made of Hydrocolloid similar to type 5 dressing, their thicknesses are the smallest among all dressings and they have large bulk density similar to type 3 dressing that renders them stiffer to resist pressure. Since 4.27kPa is considered as the reference for the breakdown pressure of capillary, the compression resistance results prove that the weft knitted spacer fabrics have better performance to resist pressure and maintain their thickness. The findings of heel interface pressure carried out reveal that 6.69kPa and 10.47kpa are the averages of heel interface pressure for those using standard hospital mattresses when their heels are at 60 degrees and 90 degrees to the mattresses respectively. In addition, 14.50kPa is the maximum average pressure for those heels at 90o on standard mattress. Since 60o and 90o are the common natural heel postures of elderly, the relief of heel interface pressure at these postures is important. The findings of the compression studies establish that weft knitted spacer fabrics have better compression resistance and resilience compared with those of pressure ulcer wound dressings. Hence, weft knitted spacer fabrics can offer a good cushioning effect to protect wounds against mechanical collision.

15.8 References

[1]   S. F. Tong, J. Yip, K. L. Yick, and M. C. W. Yuen, "Effects of different heel angles in sleep mode on heel interface pressure in the elderly," Clinical Biomechanics, Nov 2015.

[2]   D. J. Margolis, W. Bilker, J. Knauss, M. Baumgarten and B. L. Strom, "The incidence and prevalence of pressure ulcers among elderly patients in general medical practice. Annals of epidemiology," vol. 12, no. 5, pp.321-325, Jul 2002.

[3]   A. Pearson, K. Francis, B. Hodgkinson, G. Curry, "Prevalence and treatment of pressure ulcers in northern New South Wales," Australian Journal of Rural Health, vol. 8, no. 2, pp.103-110, Apr 2000.

[4]   R. Sopher, J. Nixon, E. McGinnis, A. Gefen, "The influence of foot posture, support stiffness, heel pad loading and tissue mechanical properties on biomechanical factors associated with a risk of heel ulceration," Journal of the mechanical behavior of biomedical materials, Vol. 4, no. 4, May 2011.

[5]   C. H. LYDER, "Pressure ulcer prevention and management," Jama, vol. 289, no. 2, pp223-226, Jan 2003.

[6]   C. Weller, G. Sussman, "Wound dressings update, " Journal of pharmacy practice and research, vol. 34, no. 4, pp.318-324, Dec 2006.

[7]   J. Maklebust and M. Sieggreen, "Pressure ulcers: Guidelines for prevention and management," Lippincott Williams & Wilkins, 2001.

[8]   [8] R. O. Augustine, N. A. Kalarikkal and S. A. Thomas, "Role of wound dressings in the management of chronic and acute diabetic wounds," Diabetes Mellit Hum Health Care Holist Approach Diagn Treat, pp. 273-314, Feb 2014.

[9]   G. Basal and S. A. Ilgaz, "A functional fabric for pressure ulcer prevention," Textile Research Journal, vol. 79, no. 16, pp.1415-1426, Nov 2009.

[10]   S. Ostadabbas, R. Yousefi, M. Faezipour, M. Nourani, M. Pompeo, "Pressure ulcer prevention: An efficient turning schedule for bed-bound patients," InLife Science Systems and Applications Workshop (LiSSA), 2011 IEEE/NIH, pp. 159-162, IEEE, Apr 2011.

[11]   D. R. Thomas, "Prevention and treatment of pressure ulcers: what works? what doesn't?," Cleveland Clinic journal of medicine, vol 68, no.8, pp. 704-7, Aug 2001.

[12]   S. M. Bruer, N. Powell and G. Smith, "Three-dimensionally knit spacer fabrics: a review of production techniques and applications," Journal of Textile and Apparel, Technology and Management 4, no. 4, pp. 1-31, 2005.

[13]   R. Bagherzadeh, M. Montazer, M. Latifi, M. Sheikhzadeh, and M. Sattari, "Evaluation of comfort properties of polyester knitted spacer fabrics finished with water repellent and antimicrobial agents," Fibers and Polymers, vol 8, no. 4, pp. 386-392, Jul 2007.

[14]   Y. LIU and H. Hu, "Compression property and air permeability of weft☐knitted spacer fabrics," The Journal of the Textile Institute, vol.102, no. 4, pp. 366-372, Apr 2011.

[15]   D. M. Armakan and A. Roye, "A study on the compression behavior of spacer fabrics designed for concrete applications," Fibers and Polymers, vol. 10, no. 1, pp. 116-123, Feb 2009.

[16]   J. Yip, and S. P. Ng, "Study of three-dimensional spacer fabrics: Physical and mechanical properties," Journal of materials processing technology, vol. 206, no. 1, pp359–364, Sep 2008.

[17]   M. Abounaim, G. Hoffmann, O. Diestel and C. Cherif, "Thermoplastic composite from innovative flat knitted 3D multi-layer spacer fabric using hybrid yarn and the study of 2D mechanical properties," Composites Science and Technology, vol. 70, no. 2, pp. 363-370, Feb 2010.

[18]   R. Bagherzadeh, M. Gorji, M. Latifi, P. Payvandy and L. X. Kong, "Evolution of moisture management behavior of high wicking 3D warp knitted spacer fabrics," Fibers and polymers, vol. 13, no. 4, pp. 529-534, Apr 2012.

[19]   J. Yip, and S. P. Ng, "Study of three-dimensional spacer fabrics: molding properties for intimate apparel application," journal of materials processing technology, vol. 209, no. 1, pp. 58-62, Jan 2009.

[20] S. F. Tong, J. Yip, K. L. Yick, and M. C. W. Yuen, "Exploring use of warp-knitted spacer fabric as a substitute for the absorbent layer for advanced wound dressing," Textile Research Journal, vol. 85, no. 12, pp. 1258-1268, Jul 2015.

[21] L. Onal and M. Yildirim, "Comfort properties of functional three-dimensional knitted spacer fabrics for home-textile applications," Textile Research Journal, vol. 82, no. 17, May 2012.

[22] B. J. Robinson, "The use of a hydrofibre dressing in wound management," J Wound Care, vol. 9, no. 1, pp. 32-34, Jan 2000.

[23] N. Mao N and S. J. Russell, "The thermal insulation properties of spacer fabrics with a mechanically integrated wool fiber surface," Textile Research Journal, vol. 77, no. 12, pp. 914–922, Dec 2007.

[24] G. B. Delkumburewatte and T. Dias, "Porosity and capillarity of weft knitted spacer structures," Fibers Polymer, vol. 10, no. 2, pp. 226-230, Apr 2009.

[25] A. Davies and J. T. Williams, "The Use of spacer fabrics for absorbent medical applications," Journal of Fiber Bioengineering and Informatics, vol. 1, no. 4, pp. 321-329, 2009.

[26] A. C. Tymec, B. Pieper and K. Vollman, "A comparison of two pressure-relieving devices on the prevention of heel pressure ulcers," Advances in Skin & Wound Care, vol. 10, no. 1, pp. 39, Jan 1997.

[27] Shuk FT,Joanne Y, KIT L YK, the possibility of using weft knitted spacer fabric as the wound dressing for pressure ulcer, Proceedings of ISERD International Conference, Helsinki, Finland, 5 th -6 th September 2016

Chapter 16

Modelling of warp knit spacer fabrics by using ANN

Summary

An artificial neural network (ANN) algorithm has been developed to predict the heat and moisture transfer properties of warp knitted spacer fabrics. The input of the developed models has been the structural parameters of fabrics were used as input of developed models. The findings have revealed that ANN modeling method could serve as an useful technique in prediction of transfer phenomenon of warp knitted spacer fabrics. In the prediction of heat transfer the best performance has been attributed to the topology having one hidden layer and ten neurons, after 700 epoch. In the same manner prediction of best moisture transfer occurs at topology having two hidden layers and eight neurons, after 1000 epoch. In the case of heat and moisture transfer the mean square error of testing data have been found to be 23.69 and 6.88 respectively. Also, the maximum prediction error for heat and moisture transfer has been found to be 6.32% and 9.54% respectively.

16.1 Introduction

During the past few years the research and development of technical textiles has lead to steady growth of such category of textile structures that are suited for many areas of applications. In the production of technical textiles, warp knitting technology has found a significant place despite existing fabric formation technologies or their modifications being available. Normally warp knitted spacer fabrics consist of two inter-connected outer layers. A middle interfacing layer consisting of monofilament enables the inter-connection of these outer layers [1]. Warp knitted spacer fabrics possess excellent transversal compressibility, and also have outstanding heat and moisture transfer together coupled with satisfactory air permeability. In the automotive, medical, apparels, technical and industrial end uses where comfort is considered to be the crucial aspect of choice, such properties become highly essential. There are a number of reports relating to the mechanical properties, simulation and numerical modeling of spacer fabrics. A practically feasible model that can assess porosity and capillarity of spacer fabrics used as absorbent fabric in medical applications has been designed [2,3]. The findings of a study resulted in framing of a mathematical model which could explain transfer mechanism of water vapor created by sweating from the skin to the outer surface of spacer fabric [4]. The different characteristics of spacer fabrics including low-stress mechanical properties, air permeability and thermal conductivity have been studied [5]. Investigations have shown that in the case of a spacer fabric air permeability and thermal conductivity are closely related to the fabric density [6]. Studies have been carried out on thermal comfort properties of spacer fabrics made from double jersey machine with three different dial setting and two different spacer yarns. The finding reveals that fabric weight, thermal conductivity, thermal resistivity, air permeability and relative water vapor permeability properties are considerably influenced by dial setting and the spacer yarn type. Investigations have been carried out regarding the compression behavior of warp-knitted spacer fabrics [7-10]. Work has been reported regarding the geometry modelling of such fabrics [11]. A 3D simulation of warp knitted spacer fabrics has also been proposed[12].

Artificial neural network method (ANN) is a data evaluation technique derived from intelligent technology. ANN modeling techniques are well known for the prediction of the properties of textiles depending on particular properties and also raw materials. Such techniques are normally highly accurate and also simpler than analytical models. Furthermore, with analytical modeling certain assumptions have to be taken into account, and which reduce the accuracy of proposed models.

ANN has been widely used effectively in different textile disciplines like yarn and fabric manufacture or determination of fabric properties [13-24]. Two standing examples of applicaton of ANN method of application in textile relate to the pilling tendency of wool knits and prediction of permeability of woven fabrics [13,14]. The predictive model based on dyeing of polyester fibres and its statistical comparison with yarn hairiness as well as tensile properties of cotton-covered nylon core yarns constitute yet other good examples of application of ANN modeling in the textile [15-17]. In order to predict the pilling performance of weft knitted fabrics made from wool/acrylic blended yarns artificial neural network model has been proposed [18]. A model has been developed for the prediction of tensile properties and breaking elongation of ring spun yarns respectively [21,22]. Though considerable work has been done regarding application of ANN modeling method in textiles, there has been no report on any study on prediction of warp spacer knitted fabrics properties [23-25]. Hence, efforts have been taken predict the heat and moisture transfer capacity of warp knit spacer fabrics by development of ANN models with regard to mass, thickness, porosity and fiber type of warp knit spacer fabrics.

16.2   Technical details

The following parameters of the knit spacer fabrics have been measured as per prescribed test methods [26]

a)   Mass and thickness

b)   Porosity

c)   Dynamic moisture and heat transfer

A special instrument has been developed to measure the dynamic heat and moisture transfer in fabrics. The temperature is maintained at 370C. The driving forces for the movement of moisture vapor are the temperature and vapor gradients maintained between the points where the moisture vapor emerges from the simulated skin (35°C, 90%RH) and the ambient environment controlled at 25°C and 65% RH[27].

16.3   Variation in humidity and temperature

The variation of humidity and temperature vs time are respectively depicted in figure 1. During the initial test phase, temperature and humidity in the inner and outer surfaces of the fabric have been maintained the same as that prevailing in the surrounding environment. There is a fast increase in temperature and humidity in the micro climate and also a fast increase in the temperature and humidity in the inner fabric surface due to evaporation of sweat from the skin simulator. It leads to creation of a thermal gradient between the two faces of test fabric. A balanced condition is arrived at wherein the rate of evaporation from the skin equals the rate of diffusion from the fabric by gradual decrease in the rate of heat and humidity transfer. Finally a static condition is achieved with regard to temperature and humidity. The percentages of moisture and heat can be obtaining by taking into account the heat and moisture transfer curve of samples (Figure 1).

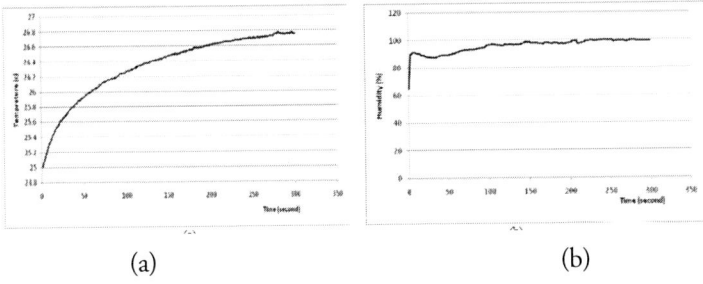

(a)   (b)

Figure 1 - Typical variation of heat and moisture transfer through fabric[34]

(a). heat transfer vs. time,

(b). moisture transfer vs. time

16.4   Modelling with ANN

Factors considered

In choosing the input parameters of ANN models, the fiber composition, fabric thickness, weight and porosity of warp knitted spacer fabrics have been considered. The moisture and heat of the fabric are critically affected by the properties of the fibre. Hence, for the purpose of training and training process of ANN models commercial fabrics produced from polyester and nylon multifilament yarns have been chosen. It has been found that the fabric thickness can affect moisture transfer capacity of fabric [28]. The permeability coefficient is directly proportional to thickness.

The heat and moisture transfer of the warp knit spacer fabrics have been considered to be the main parameters of the models. Thirty samples have been used for the training of model and five samples have been used as testing data. The specification of testing has been provided. Hence two separate model for moisture and heat transfer prediction depending on four input parameter has been proposed. Each model has 4 input units comprising of fiber composition, fabric thickness, weight and porosity of warp knitted spacer fabrics, and one output neuron comprising of moisture transfer and heat transfer %.

Back propagation learning algorithm method has been used for conducting the training process of ANN models based on gradient descent and momentum rate. The learning rate and momentum values have been set at 0.05 and 0.9 respectively.

Over fitting of artificial neural network models can be prevented by regularization or early stopping methods. The choice of either of such techniques depends on the population of data. Because of availability of small number of data, overfitting of the model has been prevented using regularization technique. This method is based on modification of performance function [29]. This function is used instead of usual performance functions of MSE and SSE (Sum Square Error). This performance function causes the network to have smaller weights and biases, causing smaller network response and over fitting [29] .

The normalizition of input and output data has been conducted in such a way that standard deviation of unity and average value of zero has been achieved. Elimination of the influence of various units of input and output parameters is mainly based on this preprocessing operation. The most common type of transfer fuctions is depicted in figure 2. Nonlinearity is the main feature of sigmoid and tangent hyperbolic functions. The tangent hyperbolic (Tansig) and sigmoid (Logsig) function compress the output data between [-1,1] and [0,1] respectively. Based on literatures, the performance of ANN model can be improved by applying sigmoid and tangent hyperbolic function in hidden layer(s) and linear function in output layer [30].

Each function can be predicted by the ANN model having one hidden layer and sigmoid transfer function in hidden layer and linear transfer function in output layer [31]. Parallel to obtain optimum topology of artificial neural network model, the best transfer function, and number of epoch were obtained by trail and error [32, 33]. Mean square error(MSE) of predicting testing data was considered as optimization criterian for judjment.

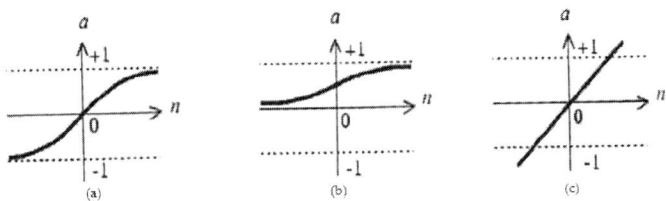

Figure 2. Schematic diagram of three used tranfer functions[34]

(a). tangent hyperbolic(tansig),
(b). sigmoid(logsig),
(c). linear

For the investigation two separate ANN model for prediction of the moisture and heat transfer capacities of warp knitted spacer fabrics have been developed. The model inputs chosen are fabric thickness, porosity, mass, and fiber composition. Each model comprises of 4 input parameters and one output unit. Since fiber composition is not quantitative parameter it has been coded as 0 and 1 for nylon and polyester fiber respectively. Also, optimization has been done for architecture parameters, number of epoch, transfer functions for hidden layer(s) and output layer, number of hidden layer(s), and number of hidden neurons.

16.5   Modeling for moisture transfer

Transfer Functions

For achieving the best transfer functions, two structure having one and two hidden layers has been chosen. Various arrangements of transfer functions have been taken into account. The number of eopchs for training was 1000. It is clear that applying tangent hyperbolic function in hidden layer(s) and linear function in output layer leads to the lowest prediction error of testing data. The lowest value of prediction error on testing data was 31.06 and relates to a network having 4-5-5-1 structure.

16.6   Number of Epoch

The number of training epoch affects the performance of training process the most. Over fitting of artificial neural network takes place with increase of epoch training epoch increases. The former also increases the duration of the training process. Hence, choice of optimum number of epoch is preferable. Thus one topology having (4-5-5-1) structure and tangent hyperbolic and linear transfer function for hidden neurons and output neuron have been chosen respectively. The number of epoch was changed from 500 to 1500 at 100 step. The findings revealed that best prediction performance of artificial neural network has been achieved after 1000 epoch.

16.7   Structure of Artificial Neural Network Model

Earlier research workers have revealed that neural networks having one hidden layer are appropriate for most of the applications. But, the second hidden layer can improve the performance of the network in the case where the input and output parameters have a complex relation. Normally trial and error method has been used to adjust the number of hidden neurons and the number of hidden layers. The optimum structure of artificial neural network has been fully based on complexity of target process as well as the relation between input and output factors. For obtaining the best topology and for assessment of ANN algorithm in modeling, twelve structure has been chosen. Based on earlier findings, number of epoch has been chosen as 1000 and tangent hyperbolic and linear transfer functions have been used for hidden and output neurons respectively. Initial assessments have shown that networks having more than two hidden layers are inappropriate for training. Trained networks have been presented having various sets of data. The mean square error of test and train data has been measured. The findings reveal that ANN model having one hidden layer and ten neurons offered the least mean square error (MSE) on testing data.

16.8   Modeling of heat transfer

Transfer Functions

For achieving the best transfer function of hidden layer(s) and output neuron(s), structures having one and two hidden layers have been chosen. Various transfer functions including linear, sigmoid, and tangent hyperbolic have been chosen for neuron(s) of hidden and output layer(s). The number of epochs has been 1000. In the case of the warp knit spacer fabrics the findings have revealed that sigmoid and linear transfer functions offer the best prediction power of heat transfer capacity. The lowest value of error on testing data was 20.10 for an network with 4-7-1 structure.

16.9 Number of Epoch

For achieving the optimum number of epochs, the factor has been changed between 500 to 1500 at 100 step. ANN model consisted of two hidden layers having five neurons in each of them. Based on earlier findings, the transfer functions of hidden layers and output neuron has been sigmoid and linear respectively. The findings revealed that the best performance of ANN model has been achieved at 700 epoch of training.

16.10 Structure of Artificial Neural Network Model

Twelve different structures with different numbers of hidden layers and number of neurons have been selected so as to identify the best structure of ANN model. On comparing the results it has been found that structure having two hidden layers and eight neurons in first and second hidden layer offered the lowest prediction error on testing data. In predicting the heat transfer of testing data, the mean square error(MSE) was 6.88.

16.11 The findings

The experimental as well as predicted values of moisture transfer of testing data accompained with prediction errors have been determined. It has been found that the maximum and minimum of prediction error are 6.32% and 4.02 % respectively. The average of prediction error is 5.01%. Such findings confirm the ability of ANN technique in prediction of moisture transfer of warp knitted spacer fabrics. The performance of optimized ANN model in prediction of heat transfer capacity of warp knitted spacer fabrics has been determined. The findings reveal that average of prediction error is 6.36% and the maximum and minimum of prediction error are 9.54 % and 3.60% respectively. Again the predictive power of artificial neural network model in prediction of moisture capacity of warp knitted fabrics is confirmed.

16.12 REFERENCES

[1]   Liu, Y., Sun, B., Hu, H., Gu, B., "Dynamic Response of 3D Biaxial Spacer Weft- knitted Composite under Transverse Impact", J. of Reinforced Plastics and Composites, Vol. 25, No. 15, 2006, pp.1629-1641.

[2]   Davies, A. and William, J., "The use of spacer fabrics for absorbent medical applications", J. of Fiber Bioengineering and informatics, Vol. 1, No. 4, 2009, pp.321-330.

[3]   Delkumbureatte, G. and Dias T., "porosity and capillary of weft knitted spacer structure", Fiber and polymers, Vol. 10, No. 2, 2009, pp.226-230.

[4]   Borhani, S., Seirafianpour, S., Hosseini, S.A., Sheikhzadeh, M., Mokhtari, M., "Computational and Experimental Investigation of Moisture Transport of Spacer Fabrics", J. of Engineered Fibers and Fabrics, Vol. 5, No. 3, 2010, pp.43-48.

[5]   Joanne, Y. and Sun-Pui, N., "Study of three-dimensional spacer fabrics: Physical and mechanical properties", J. of Materials Processing Technology, Vol. 206, No. 1-3, 2008, pp.359-364.

[6]   Ertekin, G. and Marmali, A. "Heat, Air and Water Vapor Transfer Properties of Circular Knitted Spacer Fabrics", Tekstil ve Konfeksiyon, Vol. 21, No. 4, 2011, pp.369-373.

[7]   Vasiliadis, S., Kallivertaki, A., Psilla, N., Provatidis, CH., Mecit, D., Roye, A., "Numerical modelling of the compressional behaviour of warp-knitted spacer fabrics", Fibers & Textiles in Eastern Europe, Vol. 17, No. 5, 2009, pp.56-61.

[8]   Sheikhzadeh, M., Ghane, M., Eslamian, Z., Pirzadeh, E., "A modeling study on the lateral compressive behavior of spacer fabrics", J. of the Textile Institute, Vol. 101, No. 9, 2010, pp.795-800.

[9]   Liu, Y., Hu, H., Long, H., Zhao, L.," Impact compressive behavior of warp-knitted spacer fabrics for protective applications", Textile Research J., Vol. 82, No. 8, 2012, pp.773-788.

[10]   Modeling the Moisture and Heat Transfer of Warp Knitted Spacer Fabrics Using Artificial Neural Network Algorithm 25

[11]   Mokhtari, F., Shamshirsaz, M., Latifi, M., Maroufi, M., "Compressibility behavior of warp knitted spacer fabrics based on elastic curved bar theory", J. of Engineered Fibers and Fabrics, Vol. 6, No. 4, 2012, pp.23-33.

[12]   Renkens, W. and Kyosev, Y., "Geometry modeling of warp knitted fabrics with 3D form." Textile Research J., Vol. 81, No. 4, 2011, pp. 437-443.

[13]   Zhang, L.Z., Jiang, G.M., Miao, X.H., Conga, H.L., "Three-dimensional Computer Simulation of Warp Knitted Spacer Fabric", Fibres & Textiles in Eastern Europe, Vol. 20, No. 3, 2012, pp.56- 60.

[14]   Beltran, R., Wang, L., Wang, X., "Predicting the Pilling Tendency of Wool Knits", J. of the Textile Institute, Vol. 97, No. 2, 2006, pp.129-136.

[15]   Tokarska, M., "Neural Model of the Permeability Features of Woven Fabrics", Textile Research J., Vol. 74, No. 12, 2004, pp.1045-1048.

[16]   Nasiri, M., Shanbeh, M., Tavanai, H., "Comparison Statistical Regression, Fuzzy Regression and Artificial Neural Network Modeling Methodologies in Polyester Dying", Proceeding of International Conference on Computational intelligence for Modeling, Control and Automation, 2005, pp.505-510, Austria.

[17]   Babay, A., Cheikhrouhou, M., Vermeulen, B., Rabenasolo, B., Castelain, J.M., "Selecting the optimal neural network architecture for predicting cotton yarn hairiness", J. of the Textile Institute, Vol. 96, No. 3, 2005, pp.185-192.

[18]   Ghareaghaji, A.A., Shanbeh, M., Palhang, M.,"Analysis of Two Modeling Methodologies for Predicting the Tensile Properties of Cotton-Covered Nylon Core Yarns", Textile Research J., Vol. 77, No. 8, 2007, pp.565-571.

[19]   Hasani, H. and Shanbeh, M., "Application of multiple linear regression and artificial neural network algorithms to predict the total hand value of summer knitted T-shirts", Indian J. of Fibre & Textile Research, Vol. 35, No. 3, 2010, pp.222-227.

[20]   Rejali, M., Hasani, H., Ajeli, S., Shanbeh, M.,"Optimization and prediction of the pilling performance of weft knitted fabrics produced from wool/acrylic blended yarns", Indian J. of Fibre & Textile Research, Vol. 39, No. 1, 2014, pp.83-88.

[21]   Vadood, M., Safar Johari, M., Rahai, A., "Developing a hybrid artificial neural network-genetic algorithm model to predict resilient modulus of polypropylene/polyester fiber-reinforced asphalt concrete", J. the Textile Institute, online printed, 2014, pp.1-12.

[22]   Ramesh, M.C., Rajamanickam, R., Jayaraman, S., "The prediction of yarn tensile properties by using artificial neural networks", J. the Textile Institute, Vol. 86, No. 3, 1995, pp. 459-469.

[23]   Majumdar, P.K. and Majumdar, A., "Predicting the breaking elongation of ring spun cotton yarns using mathematical, statistical, and artificial neural network models", Textile Research J., Vol. 74, No. 7, 2004, pp.652-655.

[24]   Luo, C. and Adams, D.L., "Yarn strength prediction using neural networks part I: fiber properties and yarn strength relationship", Textile Research J., Vol. 65, No. 9, 1995, pp.495-500.

[25]   Üreyen, M. and Pelin, G., "Comparison of artificial neural network and linear regression models for prediction of ring spun yarn properties. I. Prediction of yarn tensile properties", Fibers and Polymers, Vol. 9, No. 1, 2008, pp.87-91.

[26]   ASTMD3776, 1997. Standard Test Method for Thickness Textile Material, American Society for Testing and Materials.

[27]   ASTMD1777, 1997. Standard Test Method for Mass per Unit Area of Fabric, American Society for Testing and Materials.

[28]   Ziaee, M., Borhani, S., Shanbeh, M., "Evaluation of physical and mechanical properties of cotton covered polypropylene-core yarns and fabrics", Industria Textila, Vol. 62, No.1, 2011, pp. 9-13.

[29]   Wehner, J.A., Miller, B., Rebenfeld, L., "Dynamic of Water Vapor Transmission Through Barriers", Textile Research J.,Vol. 58, No. 10, 1988, pp.581-592.

[30]   Demuth, H. and Beale, M. 2001. Neural Network Toolbox/Matlab Software, the Mathworks Inc.

[31]   Gurney, K. 1997. An introduction to Neural Network, UCL Press, London.

[32]   Chattopadhyay, R., "Application of Neural Network in Yarn Manufacture", Indian J. of Fiber and Textile Research.

[33]   Ehsan G, Mohammad Z, Hossein H, and Mohsen S, Modeling the Moisture and Heat Transfer of Warp Knitted Spacer Fabrics Using Artificial Neural Network Algorithm, 2015 Textiles and Light Industrial Science and Technology, 4(1)16.

Chapter 17

Compression behavior of warp knitted spacer fabric

Summary

Studies have been carried out on the influences of test boundary condition and sample size on the compression stress-strain characteristic of a typical spacer fabric intended as a cushioning material for human body protection. Depending on the study of the load displacement curve the deformation mechanism of the fabric has been observed under a given test condition. During various phases of compression, the structure of the fabric has been observed. A number of interesting have emerged from the investigation. Only the deformation behavior of the fabric during the plateau phase in affected by the boundary condition of the compression test. Because of shear and slippage taking place in the outer layers of the fabric there are certain slight variations in the compression stress-strain curves when the test fabrics are not glued to the compression plates. The compression behavior of the fabric is evidently influenced by the size of sample, particularly in the walewise direction. But, as the sample sizein the wale direction exceeds 10 cms, the influence of the fabric size in the coursewise direction is no longer observed. It is possible to obtain reliable compression results with samples size of 10 cm × 10 cm. There are 4 different deformation phases involved in the compression behavior of the spacer fabric. A slower increase of total compression force in the initial stage is due to the loose structure of the fabric, and a linear increase of total compression force in the elastic stage is due to additional constraints by the outer layers on the endpoints of spacer monofilaments. The influence of the shortened effective lengths of spacer monofilaments as well as their torsion, shear and rotation deformations in post buckling result in the long plateau stage. During the final stage, the collapse and contacts of spacer monofilaments result in a rapidly increase of total compression force.

17.1 Introduction

Owing to properties such as good compressibility, high moisture conductivity and outstanding thermoregulation capability warp knit spacer fabric has been proposed as cushioning material for human body protection [1,2]. One of the important aspects of the fundamental mechanical properties of fabric is the compression property and relates closely to fabric handle [3]. It is the compression behavior of a warp knit spacer fabric that determines its cushioning performance. Earlier investigations have revealed that the overall compression load-displacement relationship of a warp-knitted spacer fabric can be designed to have three main stages, namely linear elasticity, plateau and densification [1, 4-6]. Such a behavior is necessary for a cushioning material to dissipate the kinetic energy of the impacting mass and at the same time maintain the maximum load below some limit [7]. But, the earlier investigations did not consider the effects of test conditions on the results. Also during the complete compression process there is no appropriate technique to simultaneously observe the deformation of spacer yarns within the fabric [1, 4, 5]. Hence, the mechanism of deformation and structure property relationships of warp knit spacer fabrics have not been properly understood due to lack of justifiable interpretation.

A warp-knitted spacer is affected by various boundary test conditions and sample sizes since it has a highly heterogeneous and discontinuous structure. There is no recognized test standard specially developed for testing the compression behavior of warp-knitted spacer fabrics. Various researchers have in their investigations have used different test methods, as per literature. As an example, contact area of 10 sq.cm at a speed of 10mm/min have been used for carrying out compression tests on spacer fabrics used for car seats. But the sample sizes have not been specified [5]. The compression properties of spacer fabrics have been tested for concrete applications by using three different contact areas as well as sample sizes in both circular and square shapeswith and without stabilization of samples at two different compression speeds [8]. Investigation has shown that the compression speed, sample size and contact area did not have significant influences on the test results. But, such conclusion holds good only for the spacer fabrics used in concrete applications. For this fabric, owing to the low areal density of outer layers the outer layers are grid nets, and the contacts between spacer yarns and outer layers as well as the contacts among spacer yarns are not evident. Also, such type of spacer fabric possesses a symmetric structure in the through-thickness as well as in-plane directions. Hence, evident influences of test boundary condition and sample size are not noticed. But, the warp-knitted spacer fabrics designed for cushioning applications differ from such type of spacer fabric. Earlier investigations have shown that they have more complex compression behavior owing to various deformation modes during compression tests, like postbuckling, shear, rotation, and contacts of spacer monofilaments among themselves and with the outer layers [2]. Such complex deformation modes can result in an unstable compression behavior of the fabrics when the sample size is below some limit. Hence it is crucial to include the influences of the test boundary condition and sample size in the analysis of the compression behavior of the spacer fabric.

Hence, the important focus of the investigation is primarily to find out how a typical warp-knitted spacer fabric for cushioning applications behaves when tested with various compression test boundary conditions and sample sizes. From here, the potential compression mechanism of the fabric is identified based on the analyses of both the compression load-displacement curve obtained under a selected test condition and the cross-sectional pictures of the fabric taken at different compression stages. This type of investigation can possibly offer certain useful information for designing warp-knitted spacer fabrics for impact protection.

17.2   Technical details

Raschel knitting machine having high speed double needle bar and six yarn guide bars has been used to produce the warp-knitted spacer fabric. The binding yarn used is polyester multifilament. For spacer yarn polyester monofilament has been used. A digital thickness tester has been used to measure the fabric thickness.

The spacer fabrics consist of two types of spacer monofilaments, i.e., vertical and inclined spacer monofilaments. On the other hand, the spacer monofilaments are knitted into the outer fabric layers to constitute the monofilament stitches simultaneously with multifilaments. Figures 1(c) and (d) depict the outer and inner surface respectively. It can be seen not only that the monofilament stitches are covered by the multifilament stitches (Figure 1(c)), but also that the spacer monofilaments are wrapped by the fluffy multifilament stitches and voids are formed in the outer fabric structure (Figure 1(d)). The above morphological identification shows that the spacer fabric has a highly heterogeneous and discontinuous structure.

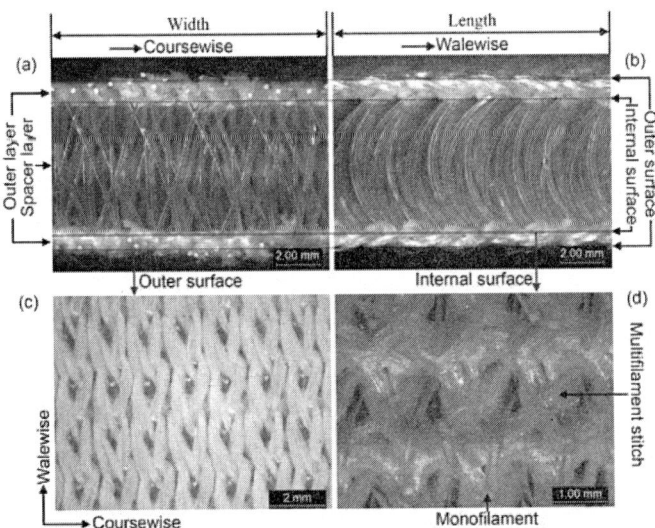

Figure 1 – Photos of spacer fabric viewed from[9]

a)   Walewise direction
b)   Coursewise direction
c)   Outer surface
d)   Internal surface

Based on the aforesaid discussions, the two critical parameters can affect the results of compression test of a warp knit spacer fabric because of the highly heterogeneous and discontinuous structure. For studying the influence of boundary condition, two cases, i.e., gluing or not gluing a fabric sample to the surfaces of two compression platens, have been taken into account. In the first case, a test spacer fabric has merely been put on the fixed compression plates without gluing it. In this case, shear can possibly take place because of possible slippage between the test fabric and the compression plates if the fabric structure is not adequately symmetrical. In the second case, two outer fabric layers have been glued to the surfaces of the two compression plates. As the in-plane movement of the two outer layers of the fabric has been constrained, only the vertical displacement has been permitted.

For studying the influence of size of the test fabric, it should be considered that a spacer fabric has a highly anisotropic structure and different mechanical behavior in different directions. The various size ratios between the walewise and coursewise directions can create various compression behavior of the fabric. Also, in the event of compression the spacer monofilaments in the fabric can contact each other. As such contacting points are based on the size of the fabric, various sizes of test fabric will also result in various compression test results of the fabric. Considering these aspects, various test spacer fabrics in rectangular shapes having various lengths and widths have been chosen. The length and width of a test fabric are defined as the length of the fabric in the walewise direction and the width of the fabric in the coursewise direction, respectively (figure 1). Two cases have been taken into account

   a)   Maintaining constant width and altering the length, and
   b)   Maintaining constant length and altering the width.

Five different lengths have been chosen in walewise and coursewise directions for the compression tests.

Instron machine has been used to conduct the compression tests based on ASTM standards.

17.3   Influence of test boundary condition

Figures 2a and b depict the compression stress-strain curves of the spacer fabric tested with two test boundary conditions under which two outer layers of the fabric have been glued or not glued to the surfaces of the compression plates, respectively. The overall trends of the stress-strain curves under the two conditions observed to be similar. The only difference in the stress-strain curves between the two boundary conditions is that the curves of the test fabrics not glued to the plates show some slight fluctuations in the plateau stage, particularly during the end of the plateau stage. It is because the various boundary conditions can lead to various actions of the fabric outer layers on the spacer monofilaments. When a fabric sample was simply placed on the smooth platens without gluing it to them, the highly buckled spacer monofilaments in the plateau stage can more easily cause the deformation and slippage of the multifilament stitches in the outer layers, resulting in the change of the constraints provided by the outer layers to the spacer monofilaments. Thus, owing to the loss of structural stability the fluctuations of the stress-strain curves have been produced in the plateau stage. Whereas, in the case where the test fabrics have been glued to the surfaces of the two compression plates, the multifilament stitches within the outer layers were not easy to deform and slip during the compression tests. The stress strain curves become smoother and more consistent since the constraints on the spacer monofilaments offered by the multifilament stitches cannot change abruptly. Such findings show that the various boundary states can affect the compression behavior of a warp-knitted spacer fabric to a certain degree.

Based on the need of a cushioning material, the two critical factors required to be optimized for a warp knit spacer fabric to fulfil such a need are the amount of the energy absorbed before attaining the densification stage and the stress level in the plateau stage. Although the slight stress fluctuations can be produced in the stress-strain curves when the test fabrics are not fixed with the compression plates, such fluctuations cannot considerably affect the cushioning performance of the fabric. In order to protect the human body from impact the spacer fabrics are usually integrated or inserted into protective clothing or equipment. Taking into account the outer layers of the spacer fabric enclosed in an impact protector cannot be securely constrained, the test condition under which a fabric is simply placed on the compression platens is nearer to a real boundary state of a spacer fabric in use. As such boundary state also renders compression tests easier to be conducted, it has been used to test fabric samples having various sizes.

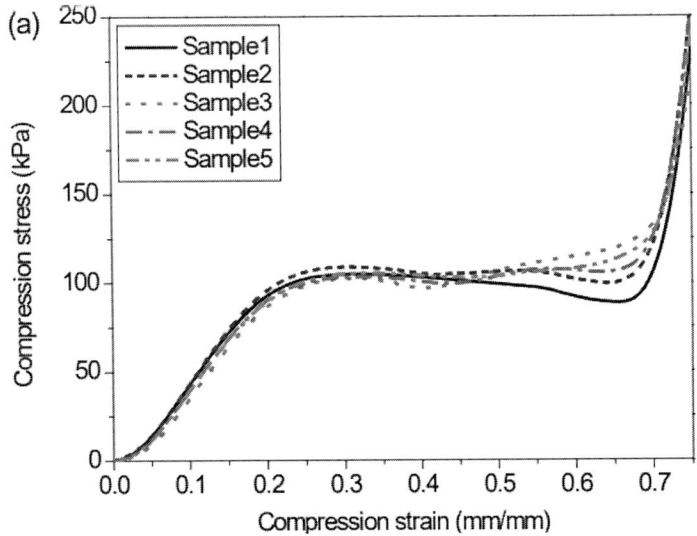

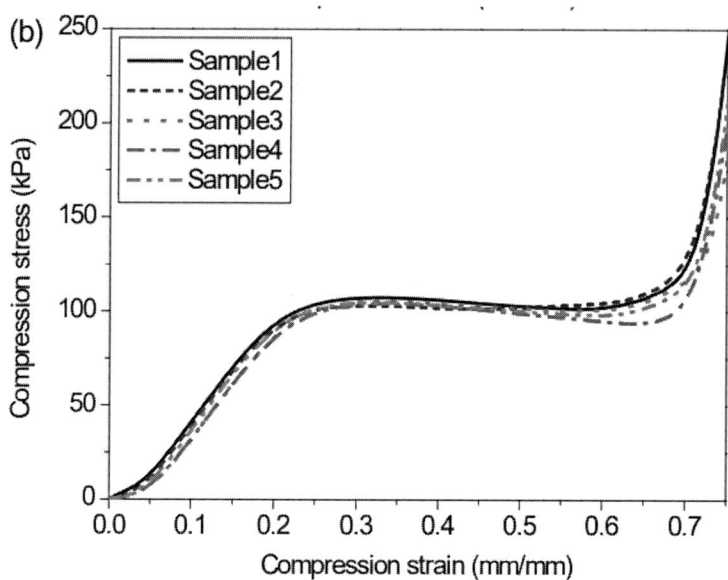

Figure 2 - Stress-strain curves of the spacer fabric under different test conditions: [9]

a)   (a) Unglued and
b)   (b) glued to platens surfaces.

17.4   Influence of size of sample

For studying the influence of sample size, a specified number of samples having various lengths and widths have been tested under the test condition not gluing them to the compression platens. Based on the European Standard "Flexible cellular polymeric materials - Determination of stress-strain characteristics in compression", the compression stress-strain characteristic (CC) expressed in kilopascals, at a defined strain, is given in the form of an equation. The parameters considered are the compression stress-strain characteristic at a particular strain, the force in newtons at the specified strain, the area in square millimeters of the tested sample. Stresses are usually specified at compressions of (25 ± 1) %, (40 ± 1) %, (50 ± 1) % and (65 ± 1) %, , and are suitably designated respectively. In the case of stress specified at compression of (25 ± 1) % is suitable to represent the plateau stress of the spacer fabric (Figure 2), and hence has been used as an indicator for assessment of the influence of the sample size on the compression stress-strain characteristic of the spacer fabric. In the design of a cushioning material for a particular end use the plateau stress is an important aspect to be considered [2].

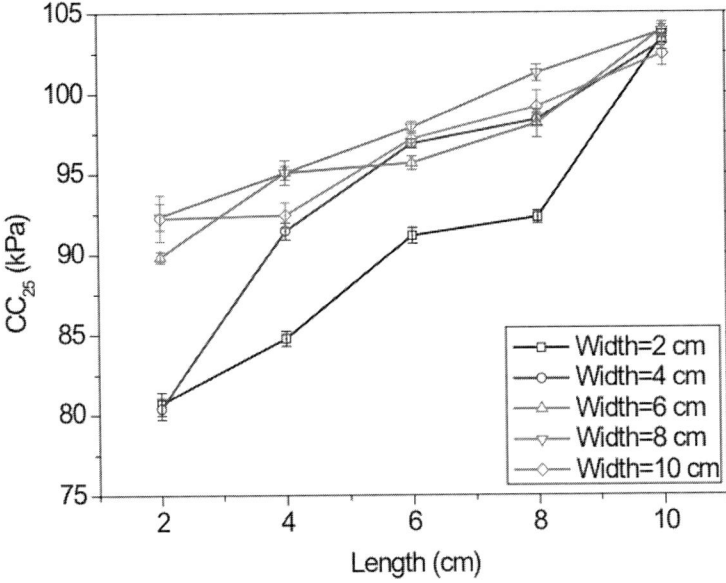

Figure3 - Influence of the sample length on test results of spacer fabric [9].

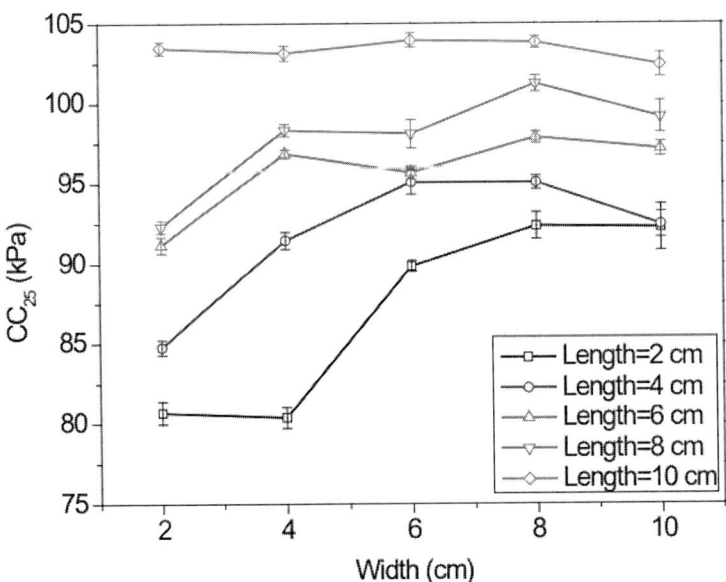

Figure 5 -Influence of the sample width on test results of spacer fabric [9].

Figures 3 and 4 depict the test results regarding the influences of the sample length and the sample width, respectively. For a given sample width the plateau stress of the fabric increases with the increase in the sample length. It implies that the sample length has a considerable influence on the test results of the fabric (Figure 3). It is seen that in the case of a given sample length, the plateau stress of the fabric increases initially followed by a reduction with the increase in the sample width (Figure 4). But, as the fabric length attains 10 cm, the influence of the sample width is no longer noticed. Such results show that the influence of the sample length along the walewise direction on the test results is more significant in comparison with that of the sample width along the coursewise direction.

Such phenomenon can be described through study of the arrangements of spacer monofilaments in the fabric structure. As depicted in Figure 1(b), all the spacer monofilaments have curved forms along the length direction (walewise). Because of the asymmetric geometric arrangement of the spacer monofilaments, shear between the two outer layers along the walewise direction can take place more easily due to small sample length. With the increase in the sample length As the increase of the sample length can decrease there is increase in the compression stress at the plateu stage due to the shear occurrence during compression. Differently, as shown in Figure 1(a), the geometric arrangement of spacer monofilaments in the width direction (coursewise) is more symmetrical than that in the length direction (walewise), because the spacer monofilaments are in vertical and symmetrically arranged in two inclined directions. Such arrangement renders the spacer fabric more stable and shear between the two outer fabric layers along the width direction cannot take place easily. The above studies show that the compression results of the fabric when tested with a relatively smaller sample length are not stable and cannot be used to properly represent the compression behavior of the fabric. The curves in Figure 4depict that the samples having a length of 10 cm can bring more stable test results irrespective of their widths. Hence, the compression load-displacement curve of the fabric sample in a size of 10 cm × 10 cm has been chosen to further study the compression deformation mechanism of the fabric, along with an analysis of the pictures of the fabric structure taken at different compression stages.

17.5   Compression Deformation Mechanism

In the design of warp knit spacer fabrics as cushioning materials for human body protection the compression deformation mechanism is considered crucial. The compression load displacement curve obtained under the test condition as well as pictures of the fabric structure taken at various compression stages have been used for explain in order to better interpret the potential deformation mechanism of the fabric.

In order to achieve better clarity, the load-displacement curve of the spacer fabric depicted in Figure 5is reproduced from the stress-strain curve of Sample 2 depicted in Figure 2(b). The process of compression can be split into four phases, namely, initial phase (phase I), linearly elastic phase (phase II), plateu phase (phase III), and densification phase (phase IV) base on the change of slopes in the curve so as enable the identification of the compression mechanism of the fabric [2]. The cross-sectional microscopic pictures of the spacer fabric having various displacements have been taken using a light microscopy.

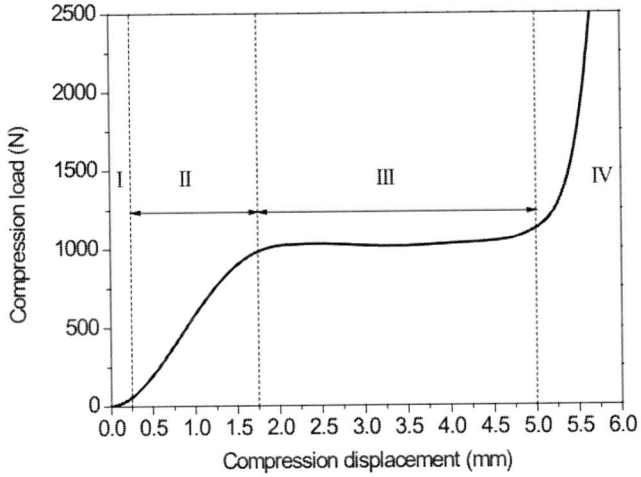

Figure5. Compression load-displacement curve of the spacer fabric [9].

During the initial phase (phase I), a lower slope has been noticed because of the compression of the loose outer layers and their ineffective constraints on the spacer monofilaments (Figure 1). A small slippage of the monofilament stitches embedded in the outer layers takes place since every loose multifilament stitch around a spacer monofilament cannot tightly constrain the spacer monofilament during this phase. On the other hand, owing to the presence of voids in the outer fabric layers the spacer monofilaments freely subject themselves to postbuckling. But, as the fabric is further compressed into linearly elastic phase (second phase), the compressed multifilament stitches get altered to a fastened structure. The stitches in the outer layers are contacted by the spacer monofilaments. Under such a situation, more constraints are caused at the contacting points between the spacer monofilaments and outer layers that render the spacer monofilaments not freely post-buckled. The postbuckling of spacer monofilaments at a larger scale with the additional constraints at the contacting points results in a rapid increase of the compression force, i.e., a stiffer mechanical behavior of the fabric.

During the plateu phase (third phase) an almost constant force is achieved. During this phase the deformation mechanism of the fabric is highly complex, and can possibly be influenced by postbuckling, torsion, shear, rotation and contacts of the spacer monofilaments among themselves and with the outer layers. Observation of the fabric taken in the walewise direction it shows that torsion, shear and rotation of the initially vertical spacer monofilaments take place, and result in the formation of three different states of the spacer monofilaments, i.e., left-oblique, right-oblique and vertical. Such deformations indicate that all the initially vertical spacer monofilaments do not behave in the same manner when compressed during the plateau phase. Such phenomenon can be because of the various constraints applied by the multifilament stitches to the endpoints of the spacer monofilaments and various initially spatial shapes of the vertical spacer monofilaments. Another possible reason is the contacts among the spacer monofilaments which could affect their postbuckling behavior. Observation of the fabric in the coursewise direction clearly shows the postbuckling process elaborately relating to the spacer monofilaments within the plateau phase. The line-to-surface contacts are formed between the spacer monofilaments and the wavy internal surface of the outer fabric layers. Division of a spacer monofilament into three portions, namely, two end portions and a middle portion, shows that as the displacement in the plateau phase increases, displacement in the plateau phase, the lengths of the two end portions contacted with the outer fabric layers increase and the length of the middle portion without contact is reduced. Hence, there is reduction in effective lengths of the spacer monofilaments in postbuckling. Therefore, the total reaction force of the spacer monofilaments increases based on the Euler-Bernoulli beam theory. Whereas, the torsional and shear deformations of the spacer monofilaments as observed in the walewise direction can result in a reduction of the total reaction force. Hence, during this phase of deformation a balance of both the effects maintains an almost constant overall reaction force of the fabric.

There is a rapid increase of the force because of the swift densification of the full fabric during the compression in the densification phase (phase IV).During this phase, the spacer monofilaments within the fabric collapse and contact one another, and hence high stiffness is achieved.

17.6   References

[1]   Liu, Y. P., et al., "Compression behavior of warp-knitted spacer fabrics for cushioning applications", Textile Research Journal, 82, 2012, 21-26.

[2]   Liu, Y. P., et al., "Impact compressive behavior of warp-knitted spacer fabrics for protective applications", Textile Research Journal, 82, 2012, 773-788.

[3]   Matsudaira, M., Qin, H., "Features and mechanical parameters of a fabric's compressional property", Journal of the Textile Institute, 86, 1995, 476-486.

[4]   [4] Miao, X. H., Ge, M. Q., "The compression behavior of warp knitted spacer fabric", Fibres & Textiles in Eastern Europe, 16, 2008, 90-92.

[5]   [5] Ye, X., et al., "Application of warp-knitted spacer fabrics in car seats", Journal of the Textile Institute, 98, 2007, 337-344.

[6]   [6] Mokhtari, F., et al., "Compressibility behaviour of warp knitted spacer fabrics based on elastic curved bar theory", Journal of Engineered Fibers and Fabrics, 6, 2011, 23-33.

[7]   [7] Avalle, M., Belingardi, G., Montanini, R., "Characterization of polymeric structural foams under compressive impact loading by means of energy-absorption diagram", International Journal of Impact Engineering, 25, 2001, 455-472.

[8]   [8] Mecit, D., Roye, A., "Investigation of a testing method for compression behavior of spacer fabrics designed for concrete applications", Textile Research Journal, 79, 2009, 867.

[9]   9. Yanping L, Hong H, 2014, An Experimental Study of Compression Behavior of Warp-knitted Spacer Fabric, Journal of Engineered Fibers and Fabrics 61 http://www.jeffjournal.org 9(2).

Chapter 18

Warp knit spacer fabric for car seats

Summary

Owing to the presence of spacer yarns the warp-knitted spacer fabrics exhibit compression properties that differ from those of normal fabrics. During the first compression stage they exhibit very good linear elastic compressibility and thus hold a good promise for vehicle seats. The limit of the first compression stage is influenced by the bending rigidity, structure of the spacer yarn and the fabric thickness. Considering this aspect, it is easy to design suitable warp knit spacer fabrics having appropriate compression characteristic that are suited for car seat applications. Moreover, warp-knitted spacer fabrics are better at decreasing peak pressure as compared to polyurethane foam. It is possible to decrease the pressure well by use of thicker warp-knitted spacer fabrics having suitable spacer yarn structure and material.

As limits are restricted to the thickness of warp knitted spacer fabrics, many layers may be used to achieve greater thicknesses and pressure relief. In comparison with seat cushions made from polyurethane foam, car seats having cushions of the warp-knitted spacer fabric are better at reducing peak pressures. Hence, the warp knit spacer fabric will make drivers feel more comfortable. When compared with polyurethane foams, air permeability of warp-knitted spacer fabrics is much better. The structure of the warp-knitted spacer fabrics determines its air permeability. Hence, in comparison with polyurethane foams such fabrics are more breathable substrates for car seats. When compared with polyurethane foams, warp-knitted spacer fabrics exhibit greater thermal conductivity and lower thermal resistance. Thus, under warm climatic conditions warp-knitted spacer fabrics can transfer heat more effectively away from the driver's body and therefore exhibit better thermoregulation properties in comparison with polyurethane foams. The stability on thickness of warp knitted spacer fabrics compare well with polyurethane foams. Generally speaking, car seat cushions made of warp knitted spacer fabrics can offer good mechanical support and physical comfort to the driver's body. These fabrics are stronger than polyurethane foams, can be used for longer periods of time and can even be reused with new seats covers. Therefore, these fabrics used for seats are easier to recycle than polyurethane foam ones.

18.1   Introduction

Environmental aspects need to be considered in the design and development of products. The recycling of materials has become crucial in the motorcar sector. Fabrics and carpets inside cars generally comprise of many layers of various materials, and is normally a polyester fabric laminated to a soft polyurethane foam backing by an adhesive. Such a kind of construction cannot enable disassembling, and the combination of various polymer chemistries for the fabric and backing renders recycling of the assembly very hard. Moreover, effluent emission arises from the flame bonding process, and helps to combine the different layers. Thus, the use of polyurethane foam in car interiors is harmful to the environment with regard to production as well as recycling [1]. Besides the environmental problems, the car seats of present day should satisfy many requirements. They should possess an appealing design, stringent mechanical properties for durability and provide protection to passengers in the event of an accident. The comfort aspect is also important. Such comfort should comprise the mechanical support which the seat offers to the body and good climatic conditions that are considered crucial in determining the efficiency of a car driver. Climatic comfort refers to good thermoregulation that can balance the body's energy and provide good microclimate around the human skin. Investigations have been carried out to determine the driver's rectal temperature with various kinds of car seats. In the case of car seats providing good thermoregulation, the passenger/driver can still be in the comfort zone even after many hours of driving, i.e. below 37.5oC even in warm temperatures (25oC). Whereas, some kinds of seats that provide poor climatic comfort, will make the driver feel uncomfortably warm fast, since the rectal temperature increases to 37.5oC in approximately 40 min. After 120 minutes, the rectal temperature can increase to 38.2oC, which is the tolerance limit of endurable strain for a normal person. It is therefore necessary that the driver should take a break when driving in such an uncomfortable seat as otherwise, he can meet with an accident due to impaired physical and mental stress [2]. The field of technical textiles has grown rapidly during recent years. Such fibrous materials that possess many technical applications, can also replace certain conventional materials that prove beneficial in particular end uses [3]. The warp-knit spacer fabrics conform to this category, and they are very interesting structures because of their special 3D configuration (Figure 1). They possess good compressibility and breathability and their thickness can range between 2-60 mm. A standard spacer fabric has two separate fabrics that are linked by spacer yarns, which complete the whole assembly [4] (Figure 1b). The spacer yarns generally used are monofilament yarns that help to maintain the space between the two independent fabrics and to achieve the necessary compressive properties in the thickness direction. Materials such as polyamide, polypropylene and particularly polyester are being used. Also, the two seperate fabrics can be knit into any type of mesh or plain structures so as to achieve the necessary dimensional, mechanical and comfort properties. Spacer fabrics show good resilience to compression, high bulk with relatively lightweight and very good moisture permeability for thermoregulation [5,6]. Besides, they offer good pressure relief [7,8].

18.2   Technical details

Mono filaments have been used as the spacer yarns and multifilament yarns have been used as the surface yarns. Six types of warp knitted with varying linear densities of yarns and four types of polyurethane foams have been used in the investigation [13]. A double needle bar warp knitting machine with six guide bars have been used for knitting the warp knit spacer fabrics. The surface fabrics produced is structurally depicted in figure 1.

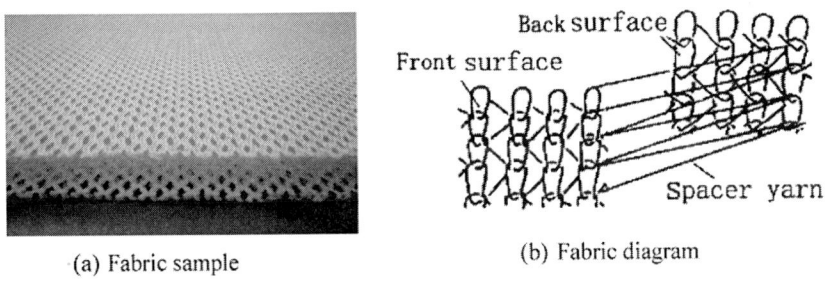

(a) Fabric sample   (b) Fabric diagram

Figure 1 – Warp knit spacer fabric [13]

The guide bars are made to go through symmetrical movements. The length of the underlaps is based on the distance between the two needle beds and on the required spacer thread angle. The following tests have been carried out on the spacer fabrics:

a)   Compression characteristics

b)   Thickness variation under dynamic and static loading

c)   Pressure relief during seating

d)   Air permeability

e)   Thermal properties

18.3   Compression characteristics

To study the compressive behaviour of different materials, samples 1, 2, 3 and A1 were chosen. Their specifications are shown in Table 1.

Table 1 – Details of materials used[13]

Material description	Wales Per cm	Courses per cm	Thickness (mm)	Mass (g/m2)	Spacer yarn angle	Type of material used
WK spacer fabric I	7.5	10	6.50	476	45o	Surface yarn 120 dtex
						PES multifilament
					90o	Spacer yarn 100 dtex
						PES monofilament
WK spacer fabric II	4.7	8	17.90	1,375	63o	Surface yarn 400 dtex
						PES multifilament
					90o	Spacer yarn 350 dtex
						PES multifilament
WK spacer fabric III	4.7	8	17.75	1,285	50o	Surface yarn 400 dtex
						PES multifilament
					90o	Spacer yarn 350 dtex
						PES monofilament
WK spacer fabric IV	4.7	8	17.54	1,100	38o	Surface yarn 400 dtex
					90o	PES monofilament
WK spacer fabric V	4.7	8	28.02	1,950	52o	Surface yarn 600 dtex
						PES multifilament

					90o	Spacer yarn 540 dtex
						PES monofilament
WK spacer fabric VI	4.7	8	43.70	2,867	52o	Surface yarn 600 dtex
						PES multifilament
					90o	Spacer yarn 540 dtex
						PES monofilament
Polyurethane foam I			28.80	576		Polyurethane foam
Polyurethane foam II			28.80	576		Polyurethane foam
Polyurethane foam III			16.16	323		Polyurethane foam
Polyurethane foam IV			14.05	442		Memory foam
Wadding			2040	458		Non woven PP wadding

WK – warpknit
PES – Polyester
PP - Polypropylene

Figure 2 depicts a typical compression/recovery curve of WK spacer fabric I. Hysteresis effect has been observed during compression and recovery as in the case of normal fabrics. Three stages are involved as depicted by this curve when compression occurs, which are related to the properties and configuration of the spacer yarn used between the two surface fabrics. During the first stage, the spacer yarn acts as an elastic spring, and links between the front and back surface of the fabrics. The yarn is subjected to axial compression, and it buckles under the compression within the elastic limit. A good bending resistance has been observed in such a case. During the second stage, the spacer yarn starts to bend above the elastic limit and hence the WK spacer fabric gets compressed more easily. At greater loads, the high-compression forces exerted on the warp-knitted spacer fabric fold the spacer yarn, and the compression of the yarn structures in the cross-sectional direction occurs, as depicted during the third stage.

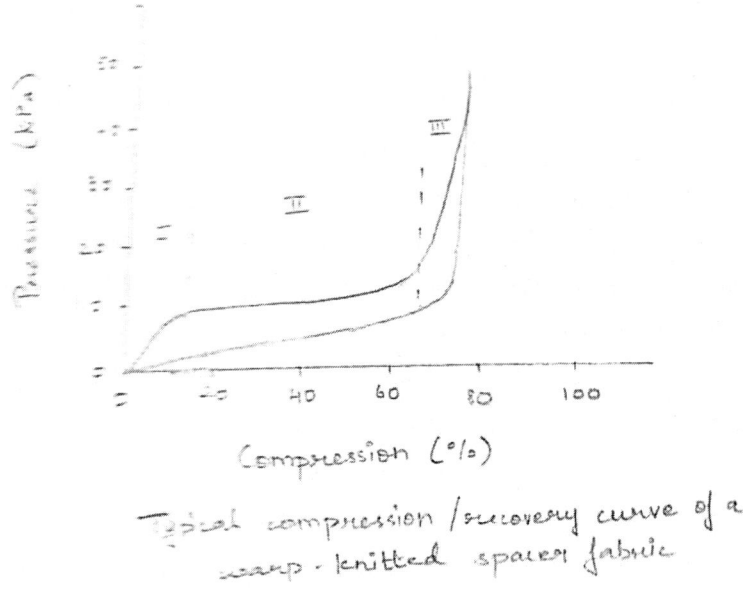

Figure 2 – Typical compression/recovery curve of a warp knitted fabric [13].

In order to achieve optimum results, the compressive load of a warp-knitted spacer fabric should be within first stage for good resilience, comfort and resistance to fabric damage. The pressure limits for the three stages differ and are based on the bending rigidity and structure of the spacer yarn and the thickness of the warp-knitted spacer fabric.

The compression curve of the warp-knitted spacer fabrics is quite different from that of the other materials listed in Table 1, as shown in Figure 3. The compression curves obtained for woven, nonwoven and knitted fabrics are same as that in the case of nonwoven wadding in Figure 3 [9-12]. The hysteresis loop of PU foam is far bigger than the ones shown for the other materials, which shows a higher dissipation of energy during the compression/recovery cycle. Upon a closer examination of Figure 5, the wadding end is found to be crushed by a pressure below 10 kPa. The PU foam and the WK spacer fabrics are strong enough to bear the weight of a person. But, the compression modulus of WK spacer fabric II is far greater in comparison with PU foam and even at 50 KPa pressure it is still in first stage of compression. The WK spacer fabric 2 in Figure 5 shows considerably higher compression rigidity in phase I than WK spacer fabric 1, since WK spacer fabric 2 is constructed with a much thicker and stiffer spacer yarn.

Various foams having the appropriate compression behavior have been used for various applications. In all cases, the compression characteristics for warp-knitted spacer fabrics is to be selected based on the end uses too. It implies that spacer fabrics should be designed for different compression behaviours (i.e. harder or softer), based on area of application.

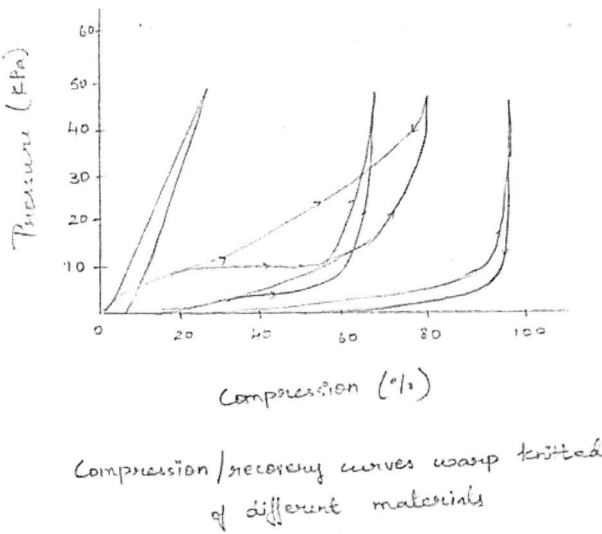

Figure 3 – Compression recovery curves warp knitted with different materials [13].

18.4   Thickness variation because of static and dynamic loading

Table 1 depicts the test fabric specifications used for the investigation of thickness variation under dynamic and static states (WK spacer fabric VI and Polyurethane foam I). Figure 4 depicts the findings. The curves depicting the behaviour of both materials follow a similar pattern, i.e. there is rapid reduction during the commencement of dynamic loading after which it stabilises. But, the initial reduction rate in polyurethane foam 1 is greater compared with that of WK knitted spacer fabric VI. After 20,000 cycles, the thickness of polyurethane foam 1 has reduced 6.6%, while the thickness of warp-knitted spacer fabric 6 has reduced 1.3%. As per earlier work, reduction in thickness takes place up to 8% in the case of some warp-knitted spacer fabrics. The reduction in thickness of WK spacer fabrics is based on the structure and properties of the spacer yarn.

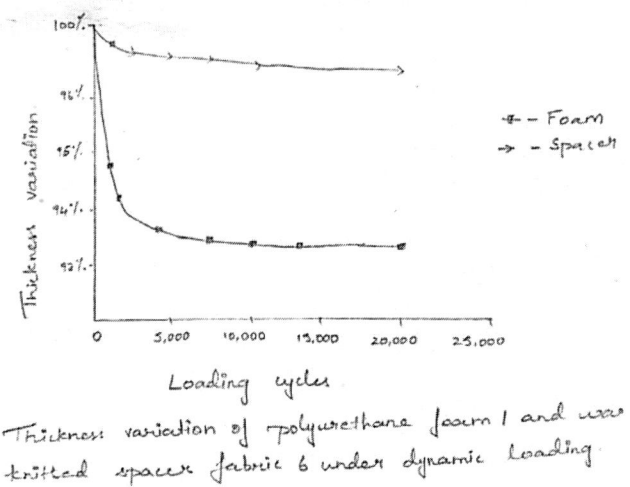

Figure 4 – Thickness variation of polyurethane foam 1 and warp knitted fabric 6under dynamic loading [13]

In the case of study of the static loading, the test fabrics have been compressed under a load of 25 kPa over a duration of 12 h, after which time the load has been removed. The thickness recovery of the test fabrics has been measured, respectively, after 15 min, 30 min, 1 h, 2 h, 5 h, 8 h and 12 h of load removal and recorded (Figure 7). The findings achieved in the case of polyurethane foam 1 as well as WK spacer fabric VI show curves of a similar shape for thickness recovery after prolonged compression. Polyurethane foam 1 exhibits a slightly better recovery rate compared with WK warp-knitted spacer fabric VI, whereas polyurethane foam 1 recovered 98.36% of the initial thickness in 12 h and WK knitted spacer fabric VI only recovered 97.98% at the same time duration.

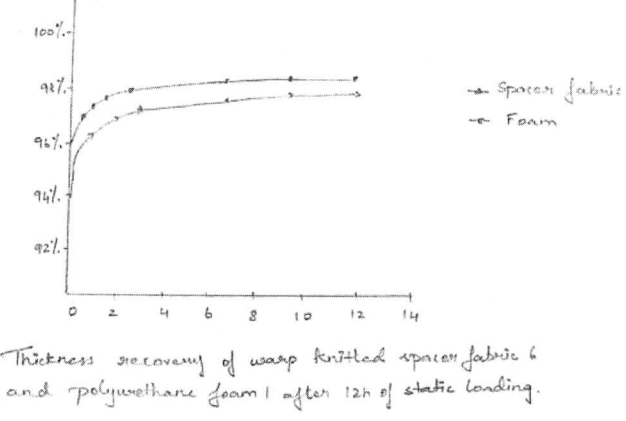

Figure 5 – Thickness recovery of warp knitted spacer fabric 6 and polyurethane foam 1 after 12 hours of static loading [13]

This very small difference of 0.38% is hardly significant. Through use of enhanced spacer yarns it is possible to improve further WK spacer fabrics resilience to compression in the thickness direction.

18.5   Pressure relief during seating

The WK spacer fabrics used for testing have been made with various fabric surface structures and thicknesses. PU foams have also chosen for the tests in order to make a comparison. The test fabric specifications have been depicted in Table 1. WK spacer fabrics I to IV A comprise of three test fabrics having the same fabric surface structure and materials but having various angles between spacer yarns and the fabric surfaces. The WK spacer fabrics V and VI are identical, except for the thicknesses that are different. With regard to PU foam III and IV, there are two types of foam. Typical 3D graphs have been depicted in Figure 6, in which the peak points on the graphs represent the peak pressure [13]. The figure depicts typical pressures at the buttocks on the seat with various cushion materials. Comparing Figures 6a and 6c, it can be seen that the three light (warm) colours disappeared in Figure 6c, implying that the high pressure has dissipated after using WK spacer fabric III. When using PU III, there is only a decrease in the area of high pressure, as depicted by comparing Figures 6 (a), (b), and (c). Comparison of these figures shows that the WK spacer fabric can decrease the pressure better than the foam.

a) without cushion

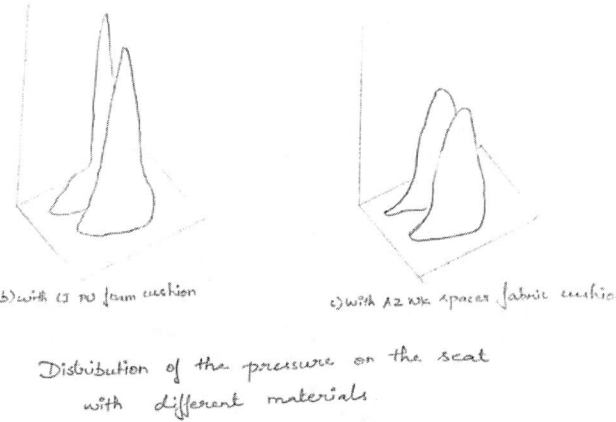

b) with CI PU foam cushion   c) with A2 WK spacer fabric cushion

Distribution of the pressure on the seat
with different materials

Figure 6 – Distribution of pressure on the seat with different materials[13]

a)   Without cushion
b)   With CI polyurethane foam cushion, and
c)   With A2 warp knitted spacer fabric cushion

Every test fabric has been measured for the peak pressure and the average pressure and has been depicted as histograms in Figure 7, where N indicates the state having no cushion; WK spacer fabrics III x2 and WK spacer fabrics III × 3 depict the states having 2 and 3 layers of WK spacer fabrics III separately.

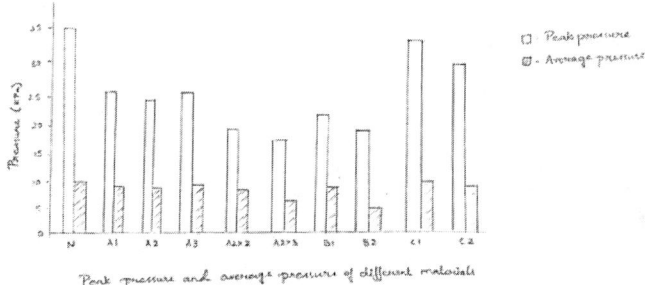

Peak pressure and average pressure of different materials

Figure 7 - Peak pressure and average pressure of different materials [13].

In the case of WK spacer fabrics II to IV, it can be observed that the angles between the spacer yarns and the fabric surfaces have an effect on pressure relief. With use of multiple layers, it can also be shown that the effectiveness of pressure relief increases with thickness.

The test fabrics WK spacer fabrics V and VI are far thicker compared with WK spacer fabrics II to IV. Figure 7 shows that WK spacer fabrics VI, which is the thickest, relieves considerably more pressure compared with any other single layer test fabric.

When number of layers in WK spacer fabrics III is increased it results better pressure relief, and 3 layers of warp knitted spacer fabric III decreases the peak pressure by 50% and the average pressure by 40%. Three layers of WK spacer fabrics III can decrease more peak pressure compared with test fabric WK spacer fabrics VI.

PU foams exhibit comparatively poor pressure relief. But, more pressure can be relieved by the use of foam with better quality and adequate compressive rigidity.

18.6 Air permeability

Air permeability of test fabrics of WK spacer fabrics II to VI and PU foams III and IV, specified in Table 1, have also been determined, as it affects comfort-related properties like breathability, which is considered crucial for cushions and mattresses. The results have been determined.

The air permeability of WK spacer fabrics are found to be far superior to PU foam for similar thicknesses. In the case of WK spacer fabrics V and VI, it can be observed that air permeability is nearly similar despite the test fabrics having quite varying thicknesses [13]. The findings in the case of WK spacer fabrics II to IV reveal that air permeability is little affected by the spacer yarn angle. It implies that air permeability is chiefly based on the cover of fabric. The difference in air permeability between PU foam samples III and IV is very large, which shows that sample IV has a less porous structure in comparison with sample IV.

18.7 Thermal properties

In order to experience comfort in seats thermal regulation is considered crucial. The test fabric specifications have been indicated in Table 1 (PU foam III and WK spacer fabric III), and the findings of the tests have been determined.

The tests have been carried out using a single layer of the WK spacer fabric (WK spacer fabric III) and PU foam (PU foam III) [13]. All the test fabrics have been maintained under the standard atmosphere condition (Temp. 23 }2°C, RH 65 }2%) for 24 h before testing. Then, the tests have been processed at similar atmospheric conditions.

The findings reveal that the WK spacer fabric exhibit greater heat conductivity and lower heat resistance compared to the PU foam. Hence, in comparison with PU foam the heat transfer from the body is easier in the case of WK spacer fabric. When a subject seats on a warp-knitted spacer fabric seat, it will be easier to prevent the build-up of heat under the buttocks comparatively to a polyurethane foam seat. This improves comfort if the driver has to drive for long periods of time under standard comfortable temperature and humidity conditions. The

thermal absorptivity of a material gives a measure of the warm/cool feeling at
initial contact with the human body. Considering this aspect, WK spacer fabrics
have a cooler feeling compared with PU foam fabrics, since thermal absorptivity
of the former is greater in all cases. It can prove beneficial when the subject
initially sits in warm climatic conditions.

18.8   References

[1]   Wilkins C, 1993, Raschel knitted spacer fabrics, Kettenwirk-Prasis, 27(3)59.

[2]   Umbach KH, 2000, Physiological comfort on car seats, Kettenwirk-Prasis, 34(1)9.

[3]   Heide M, 2000, Spacer fabric with specific protective characteristics, Melliand Textilberichte, 6, 124.

[4]   Rothe D, 2001, Warp knitted spacer fabric – design and application fields, Knitting technology, 4,14.

[5]   Schwabe D, Mohring U, and Bartels VT, 2005, Development of textiles for floor coverings and pads, Melliand Textilberichte,

[6]   Heide M, 2002, Development of functional warp knitted spacer fabrics as operating table covers, Melliand Textilberichte, 6,411.

[7]   Ye X, Fangueiro R, and Hu H, 2005, Behavior of spacer knitted fabrics used as cushions. In 4th central European conference, 7-9 september 2005, Liberec, Czech Republic.

[8]   Ye X, Hu H, and Feng X, 2005, An experimental investigation on the properties of the spacer knitted fabrics for pressure reduction, Research journal of textile and apparel, 9(3),52.

[9]   Hu J, and Newton A, 1997, Low load lateral compressional behavior of woven fabrics, Journal of textile institute, 88(3)242.

[10]   Kothari VK and Das A, 1993, The compressional behavior of spun bonded nonwoven fabrics, Journal of textile institute, 84(1)16.

[11]   Soe AK, Matsuo T, Takahashi M, and Nakajima M, 2003, Compression of plain knitted fabrics predicted from yarn properties and fabric geometry, Textile research journal, 73(10)861.

[12]   Taylor PM, and Pollet DM, 2002, Static low-load lateral compression of fabric, Textile research journal, Textile research journal, 72(11)983.

[13]   Ye X , Fangueiro X , Hu H & De Araújo M, Application of warp-knitted spacer fabrics in car seats, The Journal of The Textile Institute, 2007, 98(4)337.